Mechanical Behaviour of
Aluminium Alloys

Mechanical Behaviour of Aluminium Alloys

Special Issue Editors

Ricardo Branco
Filippo Berto
Andrei Kotousov

MDPI • Basel • Beijing • Wuhan • Barcelona • Belgrade

MDPI

Special Issue Editors
Ricardo Branco
University of Coimbra
Portugal

Filippo Berto
Norwegian University of Science and Technology
Norway

Andrei Kotousov
University of Adelaide
Australia

Editorial Office
MDPI
St. Alban-Anlage 66
Basel, Switzerland

This is a reprint of articles from the Special Issue published online in the open access journal *Applied Sciences* (ISSN 2076-3417) from 2017 to 2018 (available at: https://www.mdpi.com/journal/applsci/special_issues/Aluminium_Alloys)

For citation purposes, cite each article independently as indicated on the article page online and as indicated below:

LastName, A.A.; LastName, B.B.; LastName, C.C. Article Title. *Journal Name* **Year**, *Article Number*, Page Range.

ISBN 978-3-03897-320-1 (Pbk)
ISBN 978-3-03897-321-8 (PDF)

Cover Image Courtesy of Ricardo Branco, Berto Filippo and Andrei Kotousov.

Contents

About the Special Issue Editors vii

Ricardo Branco, Filippo Berto and Andrei Kotousov
Special Issue on "Mechanical Behaviour of Aluminium Alloys"
Reprinted from: *Appl. Sci.* **2018**, *8*, 1854, doi: 10.3390/app8101854 . . . . **1**

Elisa Fracchia, Silvia Lombardo and Mario Rosso
Case Study of a Functionally Graded Aluminum Part
Reprinted from: *Appl. Sci.* **2018**, *8*, 1113, doi: 10.3390/app8071113 . . . . **4**

Caiqi Zhao, Yangjian Zhao and Jun Ma
The Stability of New Single-Layer Combined Lattice Shell Based on Aluminum Alloy
Honeycomb Panels
Reprinted from: *Appl. Sci.* **2017**, *7*, 1150, doi: 10.3390/app7111150 . . . . **13**

Oscar Rodriguez-Alabanda, Miguel A. Narvaez, Guillermo Guerrero-Vaca and
Pablo E. Romero
Manufacturing of Non-Stick Molds from Pre-Painted Aluminum Sheets via Single
Point Incremental Forming
Reprinted from: *Appl. Sci.* **2018**, *8*, 1002, doi: 10.3390/app8061002 . . . . **25**

Torsten E. M. Staab, Paola Folgati, Iris Wolfertz and Martti J. Puska
Stability of Cu-Precipitates in Al-Cu Alloys
Reprinted from: *Appl. Sci.* **2018**, *8*, 1003, doi: 10.3390/app8061003 . . . . **36**

Ning Li, Hong Yan and Zhi-Wei Wang
Effects of Heat Treatment on the Tribological Properties of Sicp/Al-5Si-1Cu-0.5Mg Composite
Processed by Electromagnetic Stirring Method
Reprinted from: *Appl. Sci.* **2018**, *8*, 372, doi: 10.3390/app8030372 . . . . **49**

Phillip Dumitraschkewitz, Helmut Clemens, Svea Mayer, David Holec
Impact of Alloying on Stacking Fault Energies in γ-TiAl
Reprinted from: *Appl. Sci.* **2017**, *7*, 1193, doi: 10.3390/app7111193 . . . . **64**

Iban Vicario Gómez, Ester Villanueva Viteri, Jessica Montero, Mile Djurdjevic and
Gerhard Huber
The Determination of Dendrite Coherency Point Characteristics Using Three New Methods for
Aluminum Alloys
Reprinted from: *Appl. Sci.* **2018**, *8*, 1236, doi: 10.3390/app8081236 . . . . **75**

Natália Ferreira, Pedro V. Antunes, José A. M. Ferreira, José D. M. Costa and Carlos Capela
Effects of Shot-Peening and Stress Ratio on the Fatigue Crack Propagation of
AL 7475-T7351 Specimens
Reprinted from: *Appl. Sci.* **2018**, *8*, 375, doi: 10.3390/app8030375 . . . . **89**

Daniel Olvera, Gorka Urbikain, Alex Elías-Zuñiga and Luis Norberto López de Lacalle
Improving Stability Prediction in Peripheral Milling of Al7075T6
Reprinted from: *Appl. Sci.* **2018**, *8*, 1316, doi: 10.3390/app8081316 103

Zhanhui Zhang, Jiaxiang Xue, Li Jin and Wei Wu
Effect of Droplet Impingement on the Weld Profile and Grain Morphology in the Welding of
Aluminum Alloys
Reprinted from: *Appl. Sci.* **2018**, *8*, 1203, doi: 10.3390/app8071203 . . . . **114**

Chien Wan Hun, Yu-Jia Chiu, Zhiping Luo, Chien Chon Chen and Shih Hsun Chen
A New Technique for Batch Production of Tubular Anodic Aluminum Oxide Films for
Filtering Applications
Reprinted from: *Appl. Sci.* **2018**, *8*, 1055, doi: 10.3390/app8071055 . . . . **124**

About the Special Issue Editors

Ricardo Branco completed his PhD in Mechanical Engineering at the University of Coimbra. He is currently Assistant Professor at the Department of Mechanical Engineering of the University of Coimbra. His research interests include the mechanical behaviour of materials, the fatigue and fracture of engineering materials, multiaxial fatigue life prediction models, the low-cycle fatigue of high-strength steels, and numerical modelling of fatigue crack growth.

Filippo Berto completed his degree in 'Industrial Engineering' summa cum laude in 2003 at the University of Padua (Italy). After completing a PhD in Mechanical Engineering and Materials Science at the University of Florence, he worked as a researcher in the same field at the University of Padua. From 2006 to 2013, he was Assistant Professor at the University of Padua, Department of Management and Engineering, Vicenza, and from October 2014 to September 2016 he was Associate Professor. On 1 January 2016, he was appointed as International Chair of Fracture Mechanics and Fatigue at the Norwegian University of Science and Technology (NTNU).

Andrei Kotousov has actively promoted research on Applied Mechanics in Australia and internationally through a wide international collaborative network. He has participated in the investigation of reverberated structural failures in Australia and led several large industry-funded projects. Andrei serves on the Editorial Board of several journals and has chaired a number of international conferences.

applied
sciences

MDPI

Editorial

Special Issue on "Mechanical Behaviour of Aluminium Alloys"

Ricardo Branco [1],*, Filippo Berto [2] and Andrei Kotousov [3]

[1]   CEMMPRE, Department of Mechanical Engineering, University of Coimbra, Coimbra 3004-531, Portugal
[2]   Department of Industrial and Mechanical Engineering, Norwegian University of Science and Technology, 7491 Trondheim, Norway; filippo.berto@ntnu.no
[3]   School of Mechanical Engineering, University of Adelaide, Adelaide 5005, South Australia, Australia; andrei.kotousov@adelaide.edu.au
*   Correspondence: ricardo.branco@dem.uc.pt

Received: 25 September 2018; Accepted: 25 September 2018; Published: 9 October 2018

1. Introduction and Scope

Aluminium alloys are the most common type of non-ferrous material utilised for a wide range of engineering applications, namely in the automotive, aerospace, and structural industries, among others. Widespread use of these alloys in the modern world is due to a unique blend of their material properties, combining lightness, excellent strength, corrosion resistance, toughness, electrical and thermal conductivity, recyclability, and manufacturability. The last but not least important is the relatively low cost of aluminium extrusion, making aluminium alloys very attractive for applications in different key sectors of the world's economy.

Despite this great interest, extensive previous research, and knowledge accumulated in the past, recent advances in production and processing technologies combined with the development of new and more ingenious and competitive products require a profound understanding of the physical and mechanical behaviour, specifically in terms of modelling and predictions of fracture and fatigue, of aluminium alloy components. The current special issue aims to gather scientific contributions from authors working in different scientific areas, including on the improvement and modelling of mechanical properties, alloying design and manufacturing techniques, characterization of microstructure and chemical composition, as well as various advanced applications.

2. Contributions

2.1. Fabrication and Processing

Modern automotive and aerospace industries are facing new technological, economic, and environmental challenges. The development of more efficient and eco-friendly manufacturing processes is a crucial issue for commercial success in the current, highly competitive environment. This challenge has been previously explored by several authors. For example, Fracchia et al. [1] focused on the production of bi-metal engine pistons by sequential gravity-casting using both EN-AC 48000 and EN-AC 42100 aluminium alloys, and Zhao et al. [2] investigated the stability of single-layer combined lattice shells used as load-bearing structures in aerospace applications.

The mould industry has, for many years, resisted utilisation of aluminium. However, the present situation is quite different, which can be explained by the intrinsic advantages of aluminium as a moulding metal, such as low cost and faster production, to mention just a few. For example, aluminium–magnesium alloys are commonly used as moulds for the food industry. The paper written by Rodriguez-Alabanda et al. [3] proposed a novel method for the direct manufacture of non-stick moulds through a single-point incremental formation process, and studied the influence of the technological parameters on the dimensional precision of the moulds.

2.2. Characterisation and Modelling

The extremely complex relationship between the alloy design, microstructure, and mechanical properties is another area of intense research over the last few years. Below are several characteristic examples. Staab et al. [4] studied the stability of Cu-precipitates in Al–Cu alloys via the density-functional approach implemented through projected–augmented waves and plane wave expansions; the paper written by Li et al. [5] presented outcomes of a study on the effect of T6 heat treatments on tribological properties of Al–Si alloys reinforced with SiCp and processed by the electromagnetic stirring method; the paper written by Dumitraschkewitz et al. [6] provided an interesting insight into the impact of ternary-alloying elements on stacking fault energies intermetallic TiAl compounds using ab initio techniques; and finally, the paper written by Gómez et al. [7] suggested alternative methods to assess the dendrite coherency-point characteristics in $AlSi_{10}Mg$ alloys on the basis of higher-order derivatives of cooling rates and on the basis of third solid-fraction derivative curves.

Ferreira et al. [8] considered the effect of shot-peening and the stress ratio on fatigue crack propagation in aeronautical components made by the 7475-T7351 aluminum alloy, and Olvera et al. [9] analysed the stability of peripheral milling in thin-walled parts made of 7075-T6 aluminium alloys based on an enhanced multistage homotopy perturbation method.

2.3. Weldability

The weldability of aluminium alloys still represents a classic topic of research, as it did for many decades. Despite the progress made over the last few decades, many problems remain unsolved. For example, Zhang et al. [10] compared the effect of welding direction, namely vertical and inclined configurations, on the weld profile and grain morphology obtained using double pulsed-gas metal arc welding processes.

2.4. Advanced Applications

The current special issue includes a topic on high value-added products for biomedical and other demanding applications. The introduction and certification of such products is a strategic aspect in this ambit. This can be exemplified by a work by Hun et al. [11] who develop a new technique for the mass production of tubular anodic aluminium oxide films from 6061 aluminium alloy tubes, which can be applied in filtering, dialysis, and gas-diffuser processes.

Acknowledgments: The guest editors would like to thank the authors, the reviewers, and the editorial team of Applied Sciences for their valuable contributions to this special issue.

Conflicts of Interest: The authors declare no conflict of interest.

References

1. Fracchia, E.; Lombardo, S.; Rosso, M. Case Study of a Functionally Graded Aluminum Part. *Appl. Sci.* **2018**, *8*, 1113. [CrossRef]
2. Zhao, C.; Zhao, Y.; Ma, J. The Stability of New Single-Layer Combined Lattice Shell Based on Aluminum Alloy Honeycomb Panels. *Appl. Sci.* **2017**, *7*, 1150. [CrossRef]
3. Rodriguez-Alabanda, O.; Narvaez, M.A.; Guerrero-Vaca, G.; Romero, P. Manufacturing of Non-Stick Molds from Pre-Painted Aluminum Sheets via Single Point Incremental Forming. *Appl. Sci.* **2018**, *8*, 1002. [CrossRef]
4. Staab, T.; Folegati, P.; Wolfertz, I.; Puska, M. Stability of Cu-Precipitates in Al-Cu Alloys. *Appl. Sci.* **2018**, *8*, 1003. [CrossRef]
5. Li, N.; Yan, H.; Wang, Z.-W. Effects of Heat Treatment on the Tribological Properties of Sicp/Al-5Si-1Cu-0.5Mg Composite Processed by Electromagnetic Stirring Method. *Appl. Sci.* **2018**, *8*, 372. [CrossRef]
6. Dumitraschkewitz, P.; Clemens, H.; Mayer, S.; Holec, D. Impact of Alloying on Stacking Fault Energies in γ-TiAl. *Appl. Sci.* **2017**, *7*, 1193. [CrossRef]
7. Gómez, I.; Viteri, E.; Montero, J.; Djurdjevic, M.; Huber, G. The Determination of Dendrite Coherency Point Characteristics Using Three New Methods for Aluminum Alloys. *Appl. Sci.* **2018**, *8*, 1236. [CrossRef]

8. Ferreira, N.; Antunes, P.V.; Ferreira, J.A.M.; Costa, J.D.M.; Capela, C. Effects of Shot-Peening and Stress Ratio on the Fatigue Crack Propagation of AL 7475-T7351 Specimens. *Appl. Sci.* **2018**, *8*, 375. [CrossRef]

9. Olvera, D.; Urbikain, G.; Elías-Zuñiga, A.; de Lacalle, L.N.L. Improving Stability Prediction in Peripheral Milling of Al7075T6. *Appl. Sci.* **2018**, *8*, 1316. [CrossRef]

10. Zhang, Z.; Xue, J.; Jin, L.; Wu, W. Effect of Droplet Impingement on the Weld Profile and Grain Morphology in the Welding of Aluminum Alloys. *Appl. Sci.* **2018**, *8*, 1203. [CrossRef]

11. Hun, C.W.; Chiu, Y.; Luo, Z.; Chen, C.C.; Chen, S.H. A New Technique for Batch Production of Tubular Anodic Aluminum Oxide Films for Filtering Applications. *Appl. Sci.* **2018**, *8*, 1055. [CrossRef]

applied
sciences

MDPI

Article

Case Study of a Functionally Graded Aluminum Part

Elisa Fracchia [1], Silvia Lombardo [2] and Mario Rosso [3,*

[1] Department of Applied Science and Technology (DISAT), Politecnico di Torino, Viale T. Michel 5,
    15121 Alessandria, Italy; elisa.fracchia@polito.it
[2] Fonderie Officine Meccaniche Tonno (F.O.M.T) S.p.A., Viale Lidice, 8, Grugliasco, 10095 Torino, Italy;
    silvia.lombardo@fomt.it
[3] Department of Applied Science and Technology (DISAT), Politecnico di Torino,
    C.so Duca degli Abruzzi 24, 10129 Torino, Italy
*   Correspondence: mario.rosso@polito.it; Tel.: +39-011-0904-604

Received: 31 May 2018; Accepted: 5 July 2018; Published: 10 July 2018

Abstract: The growing interest in aluminum alloys is due to the excellent ductility and mechanical strength, especially in relation with their lightness. These properties make aluminum alloys one of the most used and competitive materials in the automotive sectors. In fact, at the present day, automotive components must guarantee high mechanical and thermal properties in order to ensure low emissions of the vehicle. Despite that, harsh operating conditions can lead to a rupture in aluminum components, especially if subjected to both thermal and mechanical loads. In this panorama, aluminum functionally graded materials (FGMs) could be introduced, in order to produce a single piece with different properties that fulfill all the piece requirements. In this work, considering the typical application of the aluminum alloys as engine pistons, FGMs are realized by sequential gravity casting with the piston alloy EN AB 48000 and the alloy EN AB 42100. Tensile tests on these bi-metal parts give good results in terms of mechanical strength, elongation rates and alloys bonding.

Keywords: aluminum FGM; gravity casting; sequential casting; mechanical properties

---

1. Introduction

Functionally graded materials (FGMs) are advanced engineering composite materials that exhibit a spatial gradient in composition and/or morphology with the aim to satisfy specific requirements. There are three different types of FGMs [1]: chemical composition gradient FGMs, microstructural gradient FGMs and porosity gradient FGMs. In the chemical composition gradient FGMs, the chemical composition is gradually varied along the spatial position into the material. The microstructure gradient FGMs are characterized by a microstructural variation into the same material: the microstructure is tailored with the aim to obtain the required properties in certain parts of the piece. Finally, in the porosity gradient FGMs, the porosity changes with the space position into the material, changing pore shape and/or their size. These materials found applications in a variety of field, such as aerospace, nuclear, electrical, biomedical, defense and automotive sectors [1–3].

There are several methods to obtain FGMs: gas-based methods, liquid-phase methods and solid-phase methods [2], and among all this production processes, metallic FGMs are commonly obtained with centrifugal casting, squeeze casting, gravity casting, investment casting, sintering and infiltration techniques [4,5]. Within the casting process panorama, gravity casting is one of the simplest methods to produce FGMs. This process employs a permanent mould, commonly realized in steel. The mould is opportunely coated with a protective paint and then pre-heated at a certain temperature that must be constant in order to facilitate the cast-removal. The cast production sequence involves different steps: (i) mould cleaning; (ii) casting process; (iii) cast extraction; and (iv) sprue cut. This casting process causes a certain grade of defects, intrinsically generated by the process [6], and the final properties of castings are related to defects as well as to die design and microstructures.

Parameters that could affect the casting process causing defects are: (i) temperatures (pouring, pre-heat); (ii) time (pouring, melt holding, degasification); (iii) material and thickness of the mould-coating; (iv) mould design; and (v) pouring velocity. The types of defects and their ranking in gravity casting were classified in [7] as: (i) gas/air porosity, about 45%; (ii) shrinkage, 44%; (iii) filling related problems, 38%; (iv) cracks, 33%; (v) inclusions, 28%; and (vi) metal/die interactions, 23%.

Focusing on the combination FGM-automotive applications, and considering the high presence of the aluminum alloys in this sector [8], it is certainly possible to combine the aluminum alloys with the concept of FGM [9–12]. In fact, aluminum alloys are considered as very interesting materials, because of their lightweight, especially if compared with ferrous alloys, which lead to an important decrease in fuel consumption reducing the polluting emissions, and their high specific resistance and ductility [13,14]. Particularly, FGMs in the automotive industries could be used for engine pistons [9], leaf springs [15] and many other applications. Focusing on the piston production techniques, the most employed processes are casting and forging [16–21]. In [22], Park et al. optimized the forging process to produce aluminum pistons employing aluminum powder; in the research [23], high performance pistons were produced by additive manufacturing; and in [24], the forging process was performed on the piston-wrought-alloy EN AW 4032 (AlSi12,5MgCuNi). In [9], a microstructure-gradient FGM aluminum piston was produced by centrifugal casting: the hypereutectic alloy A390 was melted and poured into a spinning mould, and the design of the die leads to obtain a strong segregation of the primary silicon particles on the head portion of the piston that leads to an increase in wear and hardness. Similarly, with the same producing process, Huang et al. [25] realized a chemical-composition-gradient FGMs with AlSi18CuMgNi alloy.

Engine pistons are commonly produced employing aluminum–silicon alloys because of their high mechanical resistance at high temperature and good fatigue behavior [16,25–32]. Particularly, the most widely used piston alloy is the EN AC 48000. However, this alloy exhibits poor ductility and low elongation at rupture that could lead to fatigue failure in the piston skirt. This can be avoided by realizing a FGM employing a more ductile alloy in the skirt and a mechanical and thermal resistant alloy in the piston crown. In this paper, a chemical-composition-gradient FGM for automotive piston was prepared by sequential gravity casting using the piston alloy (EN AC 48000) in contact with a more ductile composition (EN AC 42100). This FGM was made focusing on the process variables and their influence on the mechanical properties obtained.

2. Materials and Methods

2.1. Materials

FGM was realized by gravity casting using two different aluminum alloys: EN AC 48000 (AlSi12CuNiMg) and EN AC 42100 (AlSi7Mg0.3) with the compositions reported in Table 1.

Table 1. Alloy composition [33].

EN AC 48000 (AlSi12CuNiMg)								
Elements	Si	Fe	Cu	Mn	Mg	Ni	Zn	Ti	Al
Min (%)	10.5	-	0.8	-	0.8	0.7	-	-	Res.
Max (%)	13.5	0.7	1.5	0.35	1.5	1.3	0.35	0.25	Res.

EN AC 42100 (AlSi7Mg0.3)							
Elements	Si	Fe	Cu	Mn	Mg	Zn	Ti	Al
Min (%)	6.5	-	-	-	0.25	-	-	Res.
Max (%)	7.5	0.19	0.05	0.10	0.45	0.07	0.25	Res.

The AlSi12CuNiMg alloy, also known as a piston alloy, is commonly used to produce piston in the automotive sector. It has eutectic composition and the maximum elongation at rupture of about 1%. On the other hand, the AlSi7Mg0.3 alloy is a hypoeutectic aluminu-silicon alloy that contains a small amount of Mg and could reach 8% of elongation at rupture. Cu and Mg are added to increase the mechanical properties as a result of the precipitation strengthening after heat treatment, while Ni, added in the piston alloy, increases the high temperature resistance [34].

2.2. Methods

An aluminum FGM was obtained with the sequential gravity casting of the composition A (AlSi7Mg0.3) followed by the composition B (AlSi12CuNiMg). Overall, five types of manual castings were performed using a mould (made in heat treated steel) that permitted to obtain square bars of 25 mm × 125 mm × 15 mm. Optimization of the casting process included: (i) selection of the pouring order for the two compositions; (ii) casting temperatures of the alloys; (iii) waiting time between the sequential castings; and (iv) mould temperature.

The pouring order depends on the alloy-gap between solidus and liquidus temperatures. As previously demonstrated by DSC analysis (Differential Scanning Calorimetry) and detailed in [10,11], the hypoeutectic alloy (AlSi7Mg0.3) was the first poured composition for its larger gap of solidification (Figure 1).

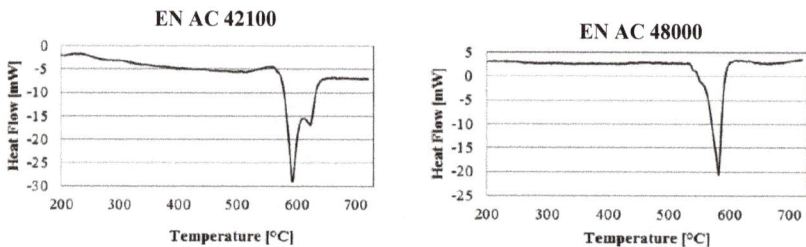

Figure 1. Differential Scanning Calorimetry (DSC) analysis [10].

The DSC curve of the hypoeutectic alloy presents two peaks: the eutectic temperature and the liquidus temperature (as can be seen in Figure 1). This gap of solidification permits to obtain a barrier effect for the second alloy (with eutectic composition) when poured, due to the solidification of the alpha-phase of the first alloy that avoids the mixing of the composition. At the same time, it is important to perform the second casting during the solidification temperature gap of the first alloy, in order to allow a metallurgical bond due to the mixing of the Si-rich phase of the second poured alloy inside the interdendritic channels of the first one.

Casting temperatures of the alloys must ensure the bonding of the two compositions avoiding the premature solidification of the first alloy poured. The waiting time between the sequential castings depends on both the temperature gap liquidus-solidus for the two alloys and the casting temperature of the alloys; in fact, this time must permit the bond of the alloys at the interfaces avoiding the mixing of the compositions. The mould temperature influences the degassing process: the higher the temperature of the mould, the faster the elimination of the humidity is.

For all these reasons, alloy temperatures were set at 710 °C (AlSi7Mg0.3) and 750 °C (AlSi12CuNiMg), with a mould temperature of 400 °C and several waiting time in the range of 10–50 s, to investigate their effect on the bonding region. No modification or refining was performed during the melting.

After the casting process, qualitatively RX analysis (X-ray radioscopic inspection) was carried out to evaluate the bonding rate, the porosity and the inclusion grade into the casted parts, especially at the interfaces between the two compositions. These observations were conducted by a Bosello High Technology Industrial X-Ray system on the casted parts.

Tensile tests were done by a ZWICK ROELL machine with the aim to evaluate the resistance of the joint between the alloys for different elapsing time. Each casting specimen was machined into the typical dog-bone shape. Considering the composite nature of the FGMs and the possibility to obtain two specimens for each FGM casted, which lead to perform a better statistical analysis of the results, it was decided to realize rectangular-shape specimens, following the standard ASTM B557-15 for samples preparation and the ASTM E8 for testing methods. After tensile tests, the fracture surfaces were observed with the scanning electron microscope (SEM), Leo 1450VP.

3. Results and Discussion

3.1. RX Analysis

RX images of the most representative samples were shown in Figure 2. In the middle of the specimens, inclusions and gas porosities that appear white were observed, as well as reported in [35], particularly in the EN AC 48000.

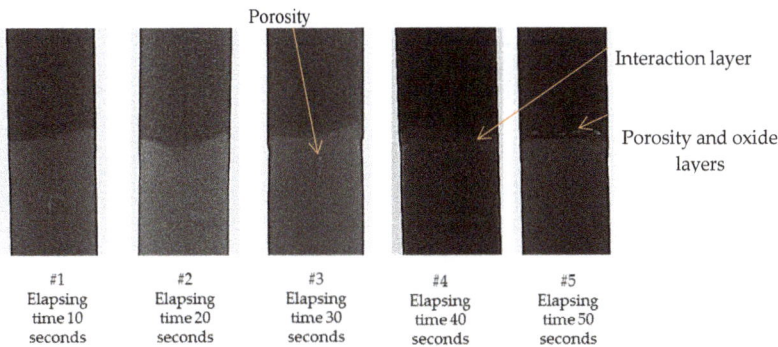

Figure 2. Sample interfaces EN AC 48000/EN AC 42100 obtained with RX analysis after sequential gravity casting at different conditions of waiting time. AlSi12CuNiMg is the dark grey part (the lower part for each image) while the other one is AlSi7Mg0.3.

After the waiting time of ten seconds, the interface appeared sharp. After twenty seconds, the solidification of the first alloy was interrupted by the casting of the second one that led to the formation of an interaction layer with the re-melting of the interface of the first poured alloy. Over thirty seconds, the interface area appeared sharp with a high grade of porosity and oxide films [36,37]. The interaction region where the two alloys met became inhomogeneous for excessive waiting time, as shown in part #5 of Figure 2.

3.2. Tensile Tests

Samples were mechanically tested for tensile resistance and results were reported in Figure 3 and in Table 2. At least three samples per each condition have been investigated. Samples with elapsing time of 40 s were not tested, while results for samples with an elapsing time of 50 s were shown in Figure 3, only to demonstrate the blatant reduction in mechanical properties.

As visible in yield strength, elongation at break and load at break increased with the increase in elapsing time until 30 s. Moreover, standard deviation decreased between elapsing time 20 and 30 s for each parameter. The tests presenting higher standard deviation had a high variability in fracture behavior of the specimens: in fact, for each elapsing time, some samples broke in correspondence of porosity and shrinkage (as shown in Figure 4A obtained with the SEM), while others broke because of oxide films (as shown in Figure 4B obtained with and Energy Dispersive X-Ray Spectroscopy EDS), causing fluctuations in both values of mechanical resistance and elongation.

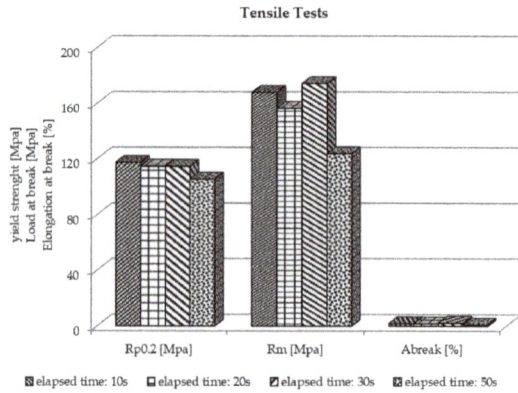

Figure 3. Mechanical properties obtained after tensile tests.

Table 2. Mechanical properties obtained with tensile tests. (E.t. = elapsing time).

Mechanical Properties		$R_{p0.2(Mpa)}$	$R_{m(Mpa)}$	A_{break} (%)
E.t. 10 s	Average	117.36	167.89	1.87
	Standard deviation	4.32	9.61	0.65
E.t. 20 s	Average	114.65	156.45	1.65
	Standard deviation	3.23	19.23	0.95
E.t. 30 s	Average	114.53	174.25	2.28
	Standard deviation	4.50	15.46	0.26
E.t. 50 s	Average	105.28	123.77	0.97
	Standard deviation	1.87	0.41	0.08

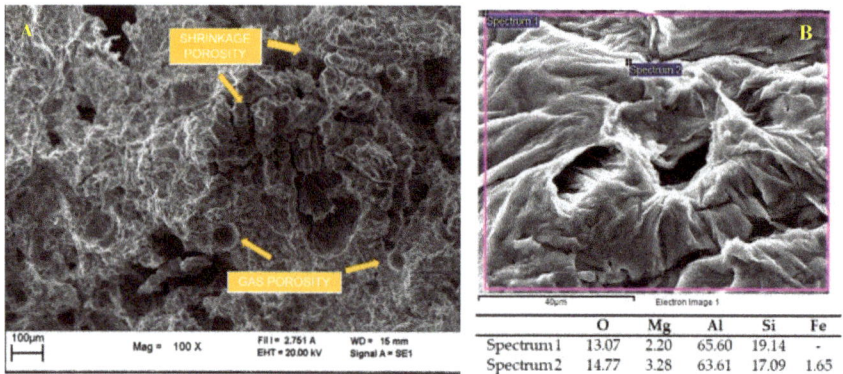

	O	Mg	Al	Si	Fe
Spectrum 1	13.07	2.20	65.60	19.14	-
Spectrum 2	14.77	3.28	63.61	17.09	1.65

Figure 4. (**A**) Sample with fracture in correspondence of gas porosity and shrinkage porosity (SEM analysis). (**B**) Sample with presence of an oxide film (Energy Dispersive X-Ray Spectroscopy EDS analysis).

Based on the obtained evaluation of the results and considering the high standard deviations, elapsing time of 30 s seemed to give the best results in terms of elongation at rupture (2.28%), while the yield strength was similar to that with the timespan 10–20 s (about 114–117 MPa). In all the cases, 50 s of elapsing time gave the worst results, demonstrating that the discontinuity caused by porosity and oxide layers seen before in the RX analysis (Figure 2) weakened the interface region between the two alloys.

3.3. Fracture Surfaces

Analysis of the fracture post-tensile tests showed various defects, especially for higher elapsing time. Shrinkage porosities, gas porosity and oxide layers were detected in more details in Table 3, which was also because of the surface turbulence associated with the filling of the mould as well as in [38].

Table 3. Surface fractures for each elapsing time.

Elapsing Time E.t.	Fracture Surfaces	Fracture Details
E.t.10s: fragile fractures in the AlSi7Mg0.3 caused by shrinkage porosity.	#1A	#1B
E.t.20s: fragile fractures in the AlSi7Mg0.3 caused by oxide and porosity.	#2A	#2B
E.t.30s: fracture in the AlSi7Mg0.3 caused by small quantity of gas porosity.	#3A	#3B
E.t.50s: fragile fractures at the interface area caused by oxide, shrinkage and gas porosity.	#4A	#4B

	O	Mg	Al	Si	Fe
Spectrum 1	50.00	10.49	37.97	1.28	0.25

As expected [10], most of the fractures happened in the weakest alloy (AlSi7Mg0.3). In just one case (E.t. 50 s), the sample reached rupture exactly into the bonding layer because of high presence of slag that weakened the meeting area.

Table 3 shows details about microstructures and EDX analysis for four surface fractures:

(#1) The sample with E.t. 10s broke in the weakest alloy, presenting shrinkage porosity in different areas of the surface (Table 3 #1B).

(#2) The sample with E.t. 20 s broke in the weakest alloy but close to the interface (in the presence of Cu and Ni in the EDS analysis). As shown in Table 3 #2B, an oxide layer was detected.

(#3) The sample with E.t. 30 s broke in the weakest alloy and presents the highest elongation. The surface is free from macro defects, as shown in Table 3 #3B.

(#4) Surface fracture of the sample with E.t. 50 s was characterized by shrinkage porosity, with typical dendrite structures, gas porosity and oxide layers (Table 3 #4A, #4B).

4. Conclusions

Sequential gravity castings of FGMs in aluminum alloys were performed. The elapsing time between the casting of the two alloys was chosen as a priority parameter in order to optimize the process obtaining effective bonding between the two compositions.

After various casting processes at different elapsing time up to 50 s, tensile tests were made to characterize the bonding behavior in each casting. Tests showed an increase in the mechanical strength after 20 s and then a decrease before 30 s. For shorter time, the alloys were mixed together in a larger and not defined area: the FGM goal was lost and the final mechanical properties were weaker. In order to maintain the alloys and their peculiar properties divided, the optimized time was between 20 and 30 s.

Certainly, the mechanical properties obtained were in line with results of previous work [10], with an increase in the mechanical resistance and the yield strength. Most of the fractures occurred in the AlSi7Mg0.3 region, which has lower mechanical resistance than AlSi12CuNiMg. The presence of the hypoeutectic alloy permitted to reach an elongation of 2%.

Overall, it appeared that only the presence of oxide and slag has caused fracture at the interface of the alloys. High elapsing time emphasizes this trend: long E.t. does not allow an effective remelting of the AlSi7Mg0.3 surface and the oxide scale, if present, remains confined into the interface area, transforming it into a weak point.

These results could be further enhanced by improving the cleanliness of the bath to remove some oxide concerns and applying a heat treatment.

Author Contributions: Conceptualization, M.R. and S.L.; methodology, S.L. and E.F.; RX analysis, S.L.; investigation and tensile tests, E.F.; data curation, E.F.; writing of the original draft preparation, E.F.; supervision, M.R.; project administration, M.R.

Funding: This research received no external funding.

References

1. Mahamood, R.M.; Akinlabi, E.T. Types of Functionally Graded Materials and Their Areas of Application. In *Mining, Metallurgy and Materials Engineering*; Springer: Cham, Switzerland, 2017.
2. Naebe, M.; Shirvanimoghaddam, K. Functionally graded materials: A review of fabrication and properties. *Appl. Mater. Today* **2016**, *5*, 223–245. [CrossRef]
3. Khan, S. Analysis of Tribological Applications of Functionally Graded Materials in Mobility Engineering. *Int. J. Sci. Eng. Res.* **2015**, *6*, 1150–1160.
4. Sobczak, J.J.; Drenchev, L. Metallic Functionally Graded Materials: A Specific Class of Advanced Composites. *J. Mater. Sci. Technol.* **2013**, *29*, 297–316. [CrossRef]
5. Singh, S.; Singh, R. Development of functionally graded material by fused deposition modelling assisted investment casting. *J. Manuf. Process.* **2016**, *24*, 38–45. [CrossRef]
6. Malhotra, V.; Kumar, Y. Study of Process Parameters of Gravity Die Casting Defects. *Int. J. Mech. Eng. Technol.* **2016**, *7*, 208–211.

7.   Bonollo, F.; Fiorese, E.; Timelli, G.; Arnberg, L.; Adamane, A.C.R. StaCast project: From a survey of European aluminium alloys foundries to new standards on defect classification and on mechanical potential of casting alloys. In Proceedings of the 71st World Foundry Congress: Advanced Sustainable Foundry, WFC 2014, Bilbao, Spain, 19–21 May 2014.
8.   Fridlyander, I.N.; Sister, V.G.; Grushko, O.E.; Berstenev, V.V.; Sheveleva, L.M.; Ivanova, L.A. Aluminum alloys: Promising materials in the automotive industry. *Met. Sci. Heat Treat.* **2002**, *44*, 365–370. [CrossRef]
9.   Arsha, A.G.; Jayakumar, E.; Rajan, T.P.D.; Antony, V.; Pai, B.C. Design and fabrication of functionally graded in-situ aluminium composites for automotive pistons. *Mater. Des.* **2015**, *88*, 1201–1209. [CrossRef]
10.   Rosso, M.; Lombardo, S.; Gobber, F. Sequential gravity casting in functionally graded aluminum alloys development. In *Light Metals*; Springer: Cham, Switzerland, 2017.
11.   Lombardo, S.; Peter, I.; Rosso, M. Gravity casting of variable composition Al alloys: Innovation and new potentialities. In Proceedings of the Aluminum Two Thousand World Congress and International Conference on Extrusion and Benchmark ICEB 2017, Verona, Italy, 20–24 June 2017.
12.   Di Ciano, M.; Caron, E.J.F.R.; Weckman, D.C.; Wells, M.A. Interface Formation During Fusion^{TM} Casting of AA3003/AA4045 Aluminum Alloy Ingots. *Metall. Mater. Trans. B Process Metall. Mater. Process. Sci.* **2015**, *46*, 2674–2691. [CrossRef]
13.   Cui, J.; Roven, H.J. Recycling of automotive aluminum. *Trans. Nonferr. Met. Soc. China* **2010**, *20*, 2057–2063. [CrossRef]
14.   Płonka, B.; Kłyszewski, A.; Senderski, J.; Lech-Grega, M. Application of Al alloys, in the form of cast billet, as stock material for the die forging in automotive industry. *Arch. Civ. Mech. Eng.* **2008**, *8*, 149–156. [CrossRef]
15.   Mehta, D.U.; Roy, D.K.; Saha, K.N. Nonlinear Analysis of Leaf Springs of Functionally Graded Materials. *Procedia Eng.* **2013**, *51*, 538–543.
16.   European Aluminium Association. *The Aluminium Automotive Manual*; European Aluminium Association: Bruxelles, Belgium, 2013; pp. 1–17.
17.   Bonollo, F.; Urban, J.; Bonatto, B.; Botter, M. Gravity and low pressure die casting of aluminium alloys: A technical and economical benchmark. *Metall. Ital.* **2005**, *5*, 23–32.
18.   Kuhlman, G.W. Forging of aluminum alloys. In *ASM Handbook, Volume 14A: Metalworking: Bulk Forming*; Semiatin, S.L., Ed.; ASM: Almere, The Netherlands, 2005; pp. 299–312.
19.   Umezawa, O.; Takagi, H.; Sekiguchi, T.; Yamashita, T.; Miyamoto, N. Novel process development with continuous casting and precise forging for AL-SI alloys to produce an engine piston. *Ceram. Trans.* **2009**, *207*, 189–200.
20.   Choi, J.I.; Park, J.H.; Kim, J.H.; Kim, S.K.; Kim, Y.H.; Lee, J.H. A study on manufacturing of aluminum automotive piston by thixoforging. *Int. J. Adv. Manuf. Technol.* **2007**, *32*, 280–287. [CrossRef]
21.   Hosokawa, H.; Higashi, K. Materials design for industrial forming process in high-strain-rate superplastic Al-Si alloy. *Mater. Res. Innov.* **2001**, *4*, 231–236. [CrossRef]
22.   Park, J.O.; Kim, K.J.; Kang, D.Y.; Lee, Y.S.; Kim, Y.H. An experimental study on the optimization of powder forging process parameters for an aluminum-alloy piston. *J. Mater. Process. Technol.* **2001**, *113*, 486–492. [CrossRef]
23.   Barbieri, S.G.; Giacopini, M.; Mangeruga, V.; Mantovani, S. A Design Strategy Based on Topology Optimization Techniques for an Additive Manufactured High Performance Engine Piston. *Procedia Manuf.* **2017**, *11*, 641–649. [CrossRef]
24.   Balducci, E.; Ceschini, L.; Morri, A.; Morri, A. EN AW-4032 T6 Piston Alloy After High-Temperature Exposure: Residual Strength and Microstructural Features. *J. Mater. Eng. Perform.* **2017**, *26*, 3802–3812. [CrossRef]
25.   Huang, X.; Liu, C.; Lv, X.; Liu, G.; Li, F. Aluminum alloy pistons reinforced with SiC fabricated by centrifugal casting. *J. Mater. Process. Technol.* **2011**, *211*, 1540–1546. [CrossRef]
26.   Nicoletto, G.; Riva, E.; di Filippo, A. High temperature fatigue behavior of eutectic Al-Si-Alloys used for piston production. *Procedia Eng.* **2014**, *74*, 157–160. [CrossRef]
27.   Dutta, S.; Kaiser, M.S. Recrystallization kinetics in Aluminum piston. *Procedia Eng.* **2014**, *90*, 188–192. [CrossRef]
28.   Mbuya, T.O.; Sinclair, I.; Moffat, A.J.; Reed, P.A.S. Micromechanisms of fatigue crack growth in cast aluminium piston alloys. *Int. J. Fatigue* **2012**, *42*, 227–237. [CrossRef]
29.   Zeren, M. The effect of heat-treatment on aluminum-based piston alloys. *Mater. Des.* **2007**, *28*, 2511–2517. [CrossRef]

30. Ma, C.; Cheng, D.; Zhu, X.; Yan, Z.; Fu, J.; Yu, J.; Liu, Z.; Yu, G.; Zheng, S. Investigation of a self-lubricating coating for diesel engine pistons, as produced by combined microarc oxidation and electrophoresis. *Wear* **2018**, *394–395*, 109–112. [CrossRef]
31. European Aluminmium Association. *Aluminium in Cars*; European Aluminmium Association: Bruxelles, Belgium, 2011; Volume 10.
32. Dyzia, M. Aluminum matrix composite (AlSi7Mg2Sr0.03/SiCp) pistons obtained by mechanical mixing method. *Materials* **2018**, *11*, 14.
33. UNI Ente Nazionale Italiano di Unificazione. *UNI EN 1706'*; UNI Ente Nazionale Italiano di Unificazione: Geneva, Switzerland, 2010; p. 30.
34. Stadler, F.; Antrekowitsch, H.; Fragner, W.; Kaufmann, H.; Pinatel, E.R.; Uggowitzer, P.J. The effect of main alloying elements on the physical properties of Al-Si foundry alloys. *Mater. Sci. Eng. A* **2013**, *560*, 481–491. [CrossRef]
35. Pirovano, R.; Mascetti, S. Tracking of collapsed bubbles during a filling simulation Die-casting. *Metall. Ital.* **2016**, *6*, 37–40.
36. Dispinar, D.; Campbell, J. Porosity, hydrogen and bifilm content in Al alloy castings. *Mater. Sci. Eng. A* **2011**, *528*, 3860–3865. [CrossRef]
37. Dispinar, D.; Akhtar, S.; Nordmark, A.; di Sabatino, M.; Arnberg, L. Degassing, hydrogen and porosity phenomena in A356. *Mater. Sci. Eng. A* **2010**, *527*, 3719–3725. [CrossRef]
38. El-Sayed, M.A.; Hassanin, H.; Essa, K. Effect of casting practice on the reliability of Al cast alloys. *Int. J. Cast Met. Res.* **2016**, *29*, 350–354. [CrossRef]

applied
sciences

MDPI

Article

The Stability of New Single-Layer Combined Lattice Shell Based on Aluminum Alloy Honeycomb Panels

Caiqi Zhao *, Yangjian Zhao and Jun Ma

Key Laboratory of Concrete and Prestressed Concrete Structure, Ministry of Education,
School of Civil Engineering, Southeast University, Nanjing 210096, China; zyj66720202@163.com (Y.Z.);
majunmajunb@163.com (J.M.)
* Correspondence: 101000815@seu.edu.cn; Tel.: +86-25-8620-5622

Received: 16 October 2017; Accepted: 6 November 2017; Published: 9 November 2017

Abstract: This article proposes a new type of single-layer combined lattice shell (NSCLS); which is based on aluminum alloy honeycomb panels. Six models with initial geometric defect were designed and precision made using numerical control equipment. The stability of these models was tested. The results showed that the stable bearing capacity of NSCLS was approximately 16% higher than that of a lattice shell with the same span without a reinforcing plate. At the same time; the properties of the NSCLS were sensitive to defects. When defects were present; its stable bearing capacity was decreased by 12.3% when compared with the defect-free model. The model with random defects following a truncated Gaussian distribution could be used to simulate the distribution of defects in the NSCLS. The average difference between the results of the nonlinear analysis and the experimental results was 5.7%. By calculating and analyzing nearly 20,000 NSCLS; the suggested values of initial geometric defect were presented. The results of this paper could provide a theoretical basis for making and revising the design codes for this new combined lattice shell structure.

Keywords: honeycomb sandwich structure; single-layer combined lattice shell; stability test; random defect mode method; defect-sensitive structure

1. Introduction

A honeycomb sandwich structure is a typical lightweight and high-strength biomimetic structure [1–4]. Based on the characteristics of a lightweight beetle forewing structure [5,6], Chen et al. developed an integrated biomimetic honeycomb sandwich structure [7,8], which has advantages, such as cementing free, single cast forming, and excellent mechanical properties [6,9]. Currently, this plate is composed of reinforced basalt fibre epoxy resin composite material; the thickness of the honeycomb wall is approximately 2 mm [10]. By comparison, the thickness of the honeycomb core wall of the aluminium alloy honeycomb sandwich structure is approximately 0.05 mm, which is approximately 1/40 of the former. The latter is more suitable for a large span spatial structure, which is especially sensitive to dead weight [11–13]. To improve the overall stability and ultimate bearing capacity of single-layer aluminum alloy lattice shells, in this paper, the authors propose a new spatial structure system made from an aluminum alloy. The system consists of lightweight high-strength aluminum alloy honeycomb panels that are reliably connected to the single-layer aluminum alloy lattice shell by special connecting parts. This forms a new type of single-layer aluminum alloy combined lattice shell structure. In this structure, which is referred to as a new spatial combined lattice shell (NSCLS), the plates and rods support and reinforce each other [14]. Our previous studies [15–17] showed that the new spatial structure has characteristics, such as a large span, light dead weight, and high total rigidity. Because the honeycomb panels support and reinforce the structure, the overall stiffness and stable bearing capacity of the NSCLS structure are obviously higher than those of a single-layer aluminum alloy lattice shell structure with the same span. Under the same load conditions, the new system can sustain a structure with a larger span. For these reasons, it has good application prospects (Figure 1).

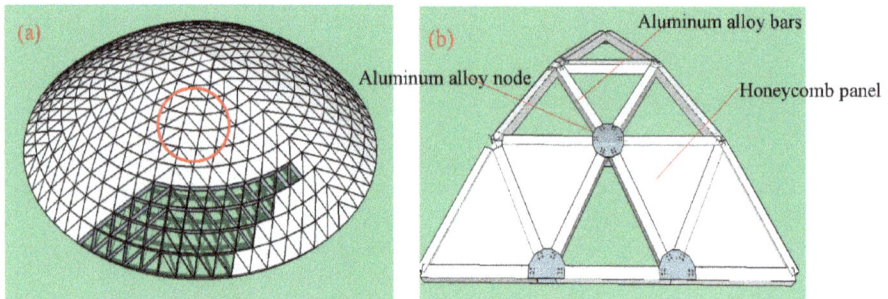

Figure 1. Schematic diagram of a large-span single-layer aluminum alloy honeycomb panel combined lattice shell: (**a**) The overall structure; (**b**) A local area, magnified.

In the study of single-layer aluminum alloy lattice shells, Hongbo Liu et al. [18] analyzed with nonlinear finite method to clarify the stability performance of aluminum alloy single-layer latticed shell. The suggested values of rise/span ratio and initial imperfection were presented. The influencing coefficients of initial imperfection and material nonlinearity on stability bearing capacity were obtained. Sugizaki et al. [19] tested four single-layer aluminum alloy spherical lattices using reduced-scale models. They studied the effects of three factors (high span ratio, load distribution pattern, and grid form) on the steady bearing capacity. The bearing capacity of the lattice shell with a mixed quadrilateral and triangular mesh was close to that of the one with an entirely triangular mesh. The influence of the nodal domain was taken into account in the finite element analysis [20]. Xiaonong Guo et al. [21] devoted to investigate the resistance of aluminum alloy gusset joint plates. It was found that the main collapse modes of AAG joint plates include the block tearing of top plates and the local buckling of bottom plates. Zechao Zhang et al. [22] analyzed the performance of aluminum alloy single-layer latticed shell by considering the combination of dead loads, live loads, snow load, wind load, temperature effect, and other seismic dynamic loads. With the geometrical nonlinear finite element method, the stability coefficient and the weak parts of the structure were obtained when considering initial imperfections and material nonlinear. Gui Guoqing et al. [23] used a numerical simulation to analyze the geometric nonlinear stability of an aluminum alloy lattice shell. They discussed the influence of the vector-to-span ratio, the load distribution, and the supporting condition on the stable bearing capacity of the lattice shell.

In the study of the honeycomb structure's stability, Attard and Hunt [24] considered the shear deformation of the panel and the core honeycomb layer. Using piecewise first-order beam theory, they analyzed the overall buckling of the honeycomb structure. Bourada M, Tounsi A et al. [11,12] studied the instability of a honeycomb core with a uniform wall thickness under uni- and bidirectional in-plane compression and out-of-plane compression. Using experimental studies and numerical simulations, A. Boudjemai, R. Amri et al. [25,26] studied the macroscopic deformation and plastic instability of a honeycomb structure with uniform wall thicknesses under unidirectional in-plane compression.

It is obvious that the existing studies focus on the aluminum alloy lattice shell structure or the honeycomb panel itself. Almost no studies of a combined aluminum alloy lattice shell structure, in which an aluminum alloy honeycomb panel reinforces the lattice shell have been conducted. There are some topics that should be addressed, such as the stability of these relatively lightweight and soft combined lattice shells and the difference between its unstable failure mode and that of ordinary single layer lattice shells that lack aluminum alloy. In this paper, the authors conducted experimental studies on the stability of this new structure to defects to investigate the structure's destabilization failure mechanism and the effect of geometric imperfections on stability. The authors also verified the reliability of the finite element method. The results of this paper could provide a theoretical basis for making and revising the design codes for this new combined lattice shell structure.

2. Experimental Study on Stability of NSCLS

2.1. The Design and Fabrication of the Test Model

To study how a honeycomb panel enhances the stable bearing capacity of a lattice shell structure, in this paper, the authors designed six single-layer aluminum alloy lattice shell models. Among the six models, three were lattice shells without honeycomb panels providing reinforcement (A_1, A_2, and A_3), and the other three were lattice shell models with reinforcement (B_1, B_2, and B_3). Figure 2a–d show the configurations of the models. The model plane was a regular hexagon. The diameter was 1400 mm, and the vector height was 250 mm (as shown in Figure 2g,h). The models were made by assembling 90 bars, 37 nodes, and 54 triangular plates together. The bars were square aluminum alloy tubes. The side length was 20 mm, and the wall thickness was 2 mm. The bars were connected by circular disc node connectors 72 mm in diameter.

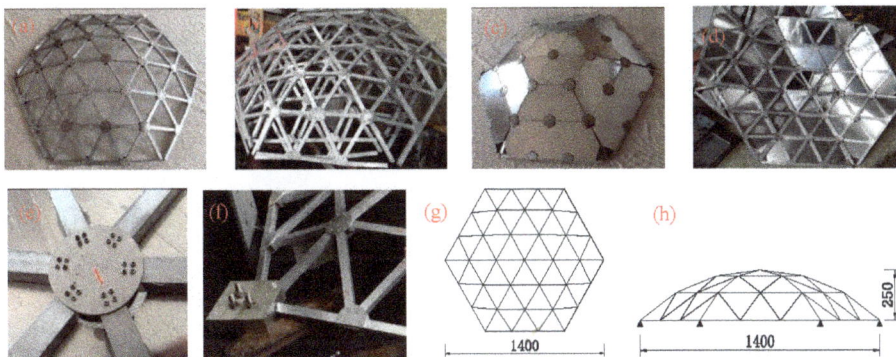

Figure 2. Overview of the test model: (**a,b**) Lattice shell model without aluminum plate (A_1, A_2, and A_3); (**c,d**) Combined lattice shell model (B1, B2, and B3); (**e**) Node in the middle; (**f**) Supporting node; (**g**) Plan view of the model; and, (**h**) Side view of the model.

To guarantee the processing accuracy of the test model, the holes on each aluminum alloy bar were precision drilled using wire cutting technology. At the upper and lower surfaces of the rods, four screw holes of 2.5 mm in diameter were machined on each surface to connect the nodal plates. To prevent the rods from colliding during assembly, the ends of the rods were cut off. Because the scale of the test model was limited by the loading equipment, the thickness of the aluminum alloy honeycomb panel was only approximately 3 mm if the components of the test model were designed at full scale. The existing processing technology was not yet able to produce such thin aluminum alloy honeycomb panels. We decided to use 1-mm-thick aluminum alloy plates in place of the honeycomb panel. Because the aluminum alloy plate and the nodal plate were placed in the same positions for nodal attachment, to accommodate the nodal plate, the corners of the aluminum plate were rounded using laser cutting equipment. The circular gussets of the connecting rods were spherical disks with a certain curvature that were stamped using a specialized mold. A total of 24 pre-machined threaded holes (their diameters were same as those of the holes at the end of the rod) were drilled on each joint plate (Figure 2e). The bolt holes at the ends of the rod were aligned with the holes in the gusset plate and tightly fit with high-strength bolts. In the model with integrated aluminum alloy plates, because the cross section of the square aluminum alloy tube was too small, the aluminum plate and the bar could not be bolted (the minimum margin requirements were not met) or welded (the thin aluminum plate was prone to burning through). Therefore, they were connected using a high-strength structural adhesive. The model was supported by fixed supports that were slotted into a computer numerical control (CNC) machine. The dimensions of the three slots were measured repeatedly. It was ensured that the actual projected length and width of the aluminum alloy square tube in the lattice model were the same, and then, the bars were placed in the slot. In this way, the rod's linear and angular displacements (Figure 2f) were well constrained. To study the influence of geometric imperfections of nodes on the bearing capacity of each model, geometric imperfections were introduced into two models (A_2 and B_2). The method was to reduce the Z coordinate of the top of the model by 5.6 mm (i.e., 1/250 of the lattice shell's span).

2.2. Loading and Testing Methods

Standard and rosette strain gages were placed on the aluminum alloy tube and the aluminum plates, respectively (Figure 3a,b). The load was provided by the MTS automatic load control system. The loading rate of 0.02 mm/s was achieved using a displacement sensor. The test site is shown in (Figure 3c).

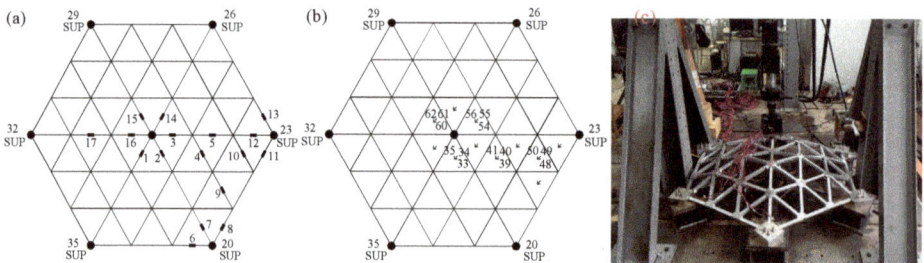

Figure 3. The strain gauge layout and the testing site: (**a**) Strain gauge locations; (**b**) Rosette strain gage locations; and, (**c**) The testing site.

2.3. Results and Discussion

2.3.1. Failure Patterns of the Test Models

Failure Pattern of the Lattice Shell Model with No Aluminum Alloy Plate

At the early stage of loading, when the load was less than 2500 N, the load increased linearly with displacement. As the load increased, it grew nonlinearly with displacement. When the displacement at the loading point reached 20~30 mm, the structure began to buckle. As the displacement continued to increase, some connecting bolts on the gusset plate broke, and the bearing capacity decreased rapidly. The bars in the area near the loading point were obviously concave. Figure 4 shows the instability failure pattern of the lattice shell structure without an aluminum alloy reinforcing plate.

Figure 4. Failure patterns of the lattice shell model without aluminum alloy plates: (**a**) The overall failure profile; (**b**) The amplified profile of a local destabilization area.

The Failure Mode of the Combined Lattice Shell Model

When the load was less than 3000 N, it increased linearly with displacement. As the load further increased, it grew nonlinearly with displacement. When the displacement at the loading point reached 15~20 mm, the structure started to buckle. As the displacement continued to increase, some connecting bolts on the gusset plate broke. A small number of connectors that were holding the aluminum alloy plate to the bar were pulled from the degumming because they could no longer sustain the load. The bearing capacity decreased rapidly. The aluminum alloy plate and the bar buckled near the loading point. The structure was deemed to have failed. Figure 5 shows the measured instability failure mode of the combined aluminum alloy lattice shell structure.

Figure 5. Failure modes of the combined lattice shell model: (**a**) The overall failure profile; (**b**) The amplified profile of a local destabilization area.

2.3.2. Load-Displacement Curve of the Test Model

Figure 6a,b show the load displacement curves of the lattice shell with no aluminum alloy reinforcing plate and combined lattice shell models, respectively. These figures show that the shapes of the curves for the six models are similar (Figure 6c) regardless of whether an aluminum alloy plate is incorporated into the lattice shell and whether there are geometric imperfections. However, the ultimate bearing capacity and the maximum displacement are different. Because the aluminum alloy plate reinforces the structure, the overall rigidity and the stable bearing capacity of the combined lattice shell structure are enhanced remarkably. The ultimate bearing capacities of the lattice shells with plates are significantly larger than that of the lattice shell with no plate. They increase by approximately 15% on average. The enhancement of the defect-free lattice shell model is more pronounced than that of the model with defects.

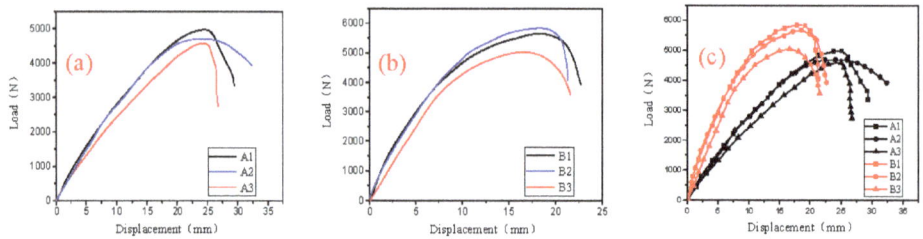

Figure 6. Load-displacement curves at the loading points of the test models: (**a**) Curves for the lattice shell models without reinforcing plates; (**b**) Curves for the combined lattice shell models; and, (**c**) Curves for the six test models.

In addition, the test results show that the sensitivity of each type of structures to geometric imperfections is different. The bearing capacity of the model with defects (A_2) is approximately 7% lower than the capacity of those without defect (A_1 and A_3). In the combined lattice shell model, the bearing capacity of the model with defects (B_2) is 12% less than the capacity of those without defects (B_1, B_3).

3. Analysis of the Nonlinear Stability of NSCLS to Defects

3.1. Basic Assumptions

In this paper, we make the following two basic assumptions when analyzing the stability of single-layer aluminum alloy lattice shells to defects using the random defect mode method. (1) The initial error in each node follows a normal probability density function within the double mean-square deviation. That means that the random variable that was representing the initial error along the X-, Y-, and Z-axes is RX/2, where the random variable X is normally distributed. R is the maximum initial error, and the error of the random variable is within [−R, R]. (2) The random variables representing the initial error of each node are independent of each other.

3.2. Analytical Methods

First, we introduce the initial error into each node in the model to create a computational model with stochastic defects; then, a nonlinear stability analysis is performed to determine the stable bearing capacity. This value is taken as a sample in the stable bearing capacity's sample space. Repeat the above steps several times to obtain multiple stable bearing capacities. In this way, a stable bearing capacity sample space containing N samples is created. Finally, based on the theory of probability and mathematical statistics, we perform a statistical analysis of the calculated steady bearing capacity of the samples to determine the final stable bearing capacity of the structure.

In the finite element analysis, Beam 188 element is simulated aluminium alloy bars and Shell 181 element is simulated aluminium alloy plates, while contacted elements are used between bars and plates. The boundary constraint condition of the model is fixed support. The material properties of aluminum alloy are as follows: elastic modulus $E = 70$ GPa, shear modulus $G = 27$ GPa, poisson's ratio $\mu = 0.33$, volume weight $\rho = 2.65 \times 10^4$ N/m^3.

3.3. Experimental Verification of Random Defect Mode Method

Before analyzing the nonlinear stability of the single-layer aluminum alloy lattice shells using the stochastic defect method, it is necessary to know the probability characteristics of each random variable, including its distribution and statistical parameters. This means that we need to determine the distribution of the initial geometric imperfections, and its mean, mean square deviation, and defect radius R (the absolute value of the maximum defect size), as well as other statistical parameters.

In this paper, we assume that the initial geometric imperfections are normally distributed. Due to the limitation of the nodal position deviation, we use a truncated Gaussian distribution. This is essentially a normal distribution, but only two truncation limits are set based on the normal distribution. Therefore, the input parameters for the distribution of the random variables that need to be determined are as follows: the mean μ, the standard deviation σ, the correlation coefficient ρ, and the truncation limits Xmin and Xmax. Because of the limited length of this paper, only two typical models of the six models are discussed (i.e., model A_1, the lattice shell with no aluminum plates or defects, and model B_2, the combined lattice shell model with defects).

3.3.1. Results for Model A_1, the Defect-Free Lattice Shell Model with No Aluminum Plate

During the analysis, the authors measured the cross-sectional dimensions at the middle and at both ends of all the aluminum alloy bars. The average of the three measured cross-sectional dimensions was used for finite element modeling. In analyzing the ultimate bearing capacity of the random defect mode, as per the relevant literature, the mean was $\mu = 0$ and the discrete coefficient was $C_d = 0.423$ (the average of the statistical results of the three coordinate components). The cutoff limit of the maximum amplitude of the deviation was $X_t = 1.4$ mm (1/1000 of the model's span), the maximum correlation coefficient was $\rho_{max} = 0.65$, and the number of random samples was N = 200.

Figure 7a,b show the history and mean square deviation of the critical load samples for model A_1, respectively. The blue curve in the middle of Figure 7b plateaus after the number of samples N exceeds 100. This indicates that the mean and mean square deviation of the critical load samples stabilize after the number of samples N exceeds 100. The red curves on the upper and lower sides are the boundaries at the 95% confidence level. The area between the curves in the graph decreases in size, which indicates that the accuracy of the history curve increases. Therefore, selecting 100 samples in practical engineering makes it possible to meet the accuracy requirements under normal conditions. The calculated results are reliable.

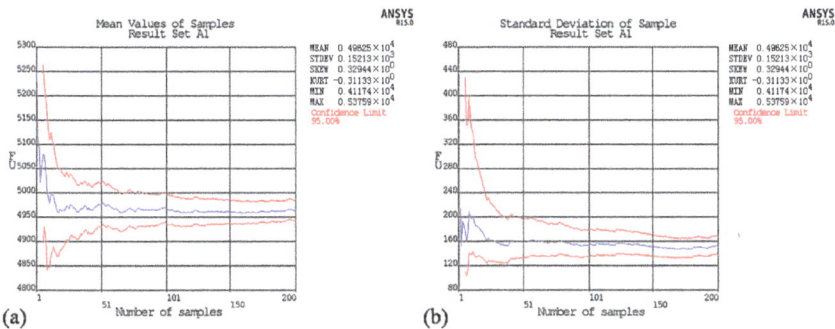

Figure 7. The mean and mean square deviation of the critical load samples for model A_1: (**a**) Mean; (**b**) Mean square deviation.

After using finite element analysis to analyze 200 samples with defects, we obtain a curve for the critical load samples (Figure 8a). Using the Gaussian method (if the random variable follows a Gaussian distribution, the probability distribution function is a straight line), we plot the probability distribution of the critical load samples (Figure 8b). The middle section of the blue curve is somewhat close to a straight line. We check that the critical load samples are normally distributed by evaluating the fit. The normal distribution assumption is acceptable at the 5% significance level.

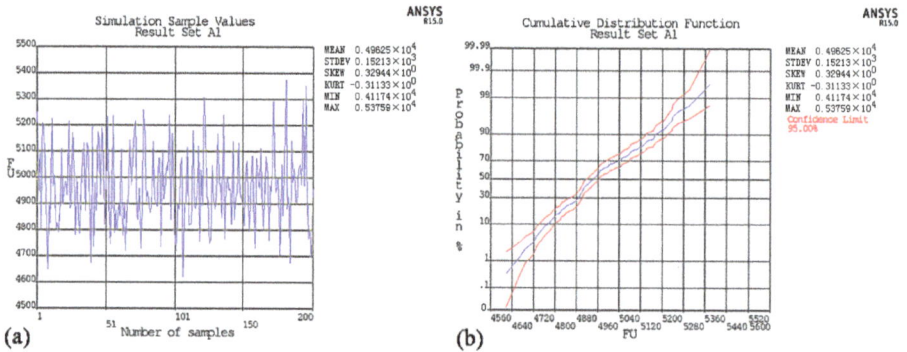

(a)

(b)

Figure 8. History and probability distribution for the critical load of model A_1: (**a**) History; (**b**) Probability distribution according to the Gaussian method.

The critical load samples are statistically analyzed and compared with the experimental results. The results are listed in Table 1. The critical loads P_{cr} in the table are determined by following the "3σ" principle (i.e., a 99.87% guarantee rate).

Table 1. Comparison of the results obtained using the random defect mode analysis and the measured values for A_1.

P_e/N	D_{max}/mm	C_d	P_{max}/N	P_{min}/N	P_μ/N	P_σ/N	P_{cr}
4879	1.4	0.423	5376	4617	4962	152	4506

In the table: P_e—The test value of the critical load; D_{max}—The maximum defect size; C_d—The discrete coefficient; P_{max}—The maximum of the critical load samples; P_{min}—The minimum of the critical load samples; P_μ—The mean of the critical load samples; P_σ—The mean square deviation of the critical load samples; P_{cr}—The standard value of the critical load (3 times the mean square deviation is accounted for).

Figure 9 shows that, according to the random defect mode analysis of the lattice shell with no aluminum plate, the predicted theoretical failure pattern (Figure 9b) is very close to the experimental failure pattern (Figure 9a).

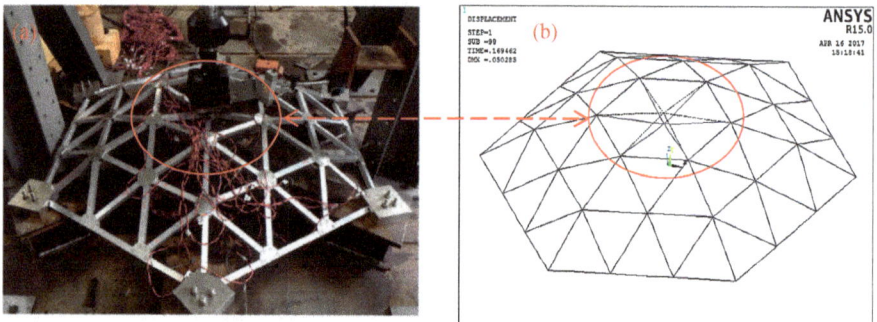

Figure 9. Comparison of the failure patterns of the test model with no plate (A_1): (**a**) Measured failure profile; (**b**) Theoretical failure profile.

3.3.2. Analysis of the Combined Lattice Shell with Defects (Model B_2)

In model B_2, the authors artificially set the geometric deviation of the node coordinates, i.e., we reduced the Z coordinate of the top of the model by 5.6 mm (1/250 of the combined lattice shell's span).

As for model A_1, we conducted a finite element analysis of 200 samples using stochastic defects to obtain curves for the critical load samples and graphed the probability distributions of the critical load samples using the Gaussian method.

The critical load sample were statistically analyzed and compared with the experimental results. The results are listed in Table 2. The critical load values P_{cr} in the table were determined according to the "3σ" principle (i.e., 99.87% guarantee rate).

Table 2. Comparison of the results obtained from the random defect mode analysis and the measured values for B_2.

P_e/N	D_{max}/mm	C_d	P_{max}/N	P_{min}/N	P_μ/N	P_σ/N	P_{cr}
5058	1.4	0.423	6210	3787	5256	220	4596

Figure 10 shows that, according to the random defect mode analysis of the combined lattice shell, the predicted theoretical failure pattern (Figure 10b) is very close to the experimental failure pattern (Figure 10a).

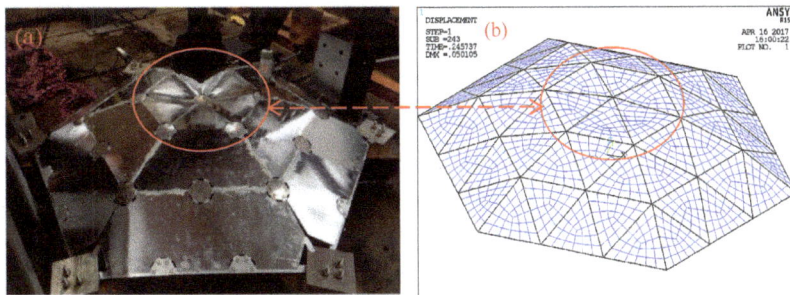

Figure 10. Comparison of the failure modes of the combined lattice shell, model B_2: (**a**) Measured failure pattern; (**b**) Theoretical failure pattern.

Table 3 shows that the results of the finite element analysis of the aluminum alloy lattice shell model obtained using the random defect mode method are very close to the experimental values. In the finite element analysis, the mean critical load and three times the mean square deviation of the standard values are accounted for. The average error of the six models is 5.7%. This can be attributed to the small discrete coefficient and defect amplitude and to the use of a truncated normal distribution for the random defects (i.e., a truncated Gaussian distribution) in this paper. It is apparent that the testing load P_e of the critical load is between P_{max} and P_{min}.

Table 3. Comparison of the critical load of the lattice shell with no plate and the combined lattice shell.

Model Configuration	Lattice Shell with No Plate			Combined Lattice Shell		
	Experimental Value (N)	Simulated Value (N)	Error (%)	Experimental Value (N)	Simulated Value (N)	Error (%)
Without defects	4879 4601	4506	7.6 2.1	5761 5556	5947	3.2 6.6
With defects	4652	4305	7.5	4958	4596	7.3

4. The Stability of NSCLS for Different Defect Sizes

The current Chinese "Technical specification for spatial lattice structures" [27] stipulates that when analyzing the stability of a lattice structure, the maximum initial geometric imperfection size (i.e., the initial deviation of the surface shape) should be set to 1/300 of the lattice shell's span. However, in this paper, the authors study a new type of single-layer aluminum alloy honeycomb combined lattice shell structure, and we find that the critical load obtained using the uniform defect mode is often not the smallest critical load. This is because the lowest-order buckling mode is frequently unable to characterize the most unfavorable distribution of defects in the structure. Therefore, the current specification of defect sizes may no longer be applicable. Therefore, it is necessary to investigate a reasonable maximum defect size for this type of structure. In this paper, we used the stochastic defect mode method to study the stable bearing capacity of a single-layer aluminum alloy lattice shell structure. To study the influence of different geometric defect sizes and plate thicknesses, the maximum defect sizes ranged from L/2000 to 16L/2000 (a total of 16 parameters), where L was the span of the combined lattice shell structure. The plate thickness was set to 0.5 mm, 1.0 mm, 1.5 mm, 2.0 mm, 2.5 mm, and 3.0 mm. The geometric dimensions and material properties that were used in the computational model were exactly the same as they were in the test model. A nonlinear finite element analysis was performed for lattice shell models with six plate thickness and 16 geometric defect sizes. Calculations were performed for 200 random samples (in total, $6 \times 16 \times 200 = 19{,}200$ samples).

By processing and analyzing the nearly 20,000 samples discussed above, the authors determined the critical load P_{cr} by following the "3σ" principle. Figure 11 shows the results. The thickness of the plate and the standard bearing capacity of the structure for different defect sizes are comprehensively accounted for.

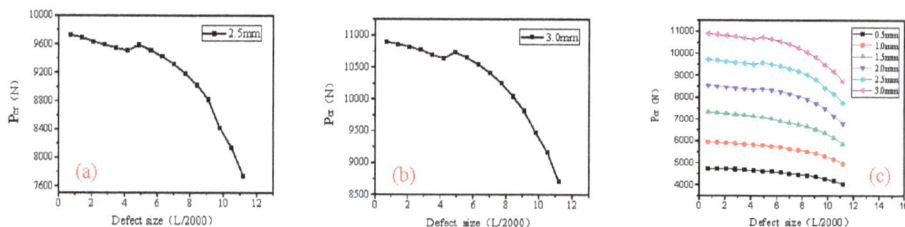

Figure 11. The critical load for different defect sizes for the lattice shell models with different plate thicknesses: (**a**) Model with a plate thickness of 2.5 mm; (**b**) Model with a plate thickness of 3.0 mm; and, (**c**) Combined lattice shell models with different plate thicknesses.

Figure 11 shows that the single-layer aluminum alloy combined lattice shell structure is sensitive to defects. As the defect size increases, the stable bearing capacity of the structure obviously decreases. When the defect size is small, i.e., below 6L/2000, the stable bearing capacity of the structure changes approximately linearly with the defect size. The bearing capacity does not change significantly. However, when the defect size exceeds 6L/2000, the stable bearing capacity of the structure decreases nonlinearly with the defect size. The decreasing slope becomes steeper as the defect size continues increasing. Therefore, the authors recommend that when designing NSCLS in the future, the maximum size of the initial geometric imperfections should be held below 3/1000 of the span of the lattice shell.

5. Conclusions

In this paper, the authors propose a new type of lightweight and high-strength aluminum alloy spatial combined lattice shell (referred to as a NSCLS). The authors analyze the structure's stability using experimental studies and a nonlinear finite element analysis. The following major conclusions are reached:

(1) By precision processing six models using a CNC machine and accounting for the initial geometric defects in the aluminum alloy lattice shell models, the authors performed a stability comparison. The results show that the overall stiffness and stable bearing capacity of the lattice shell are remarkably improved due to the aluminum alloy reinforcing plate. Regardless of whether there are geometric defects, the steady bearing capacity of the new combined lattice shells is approximately 16% higher than that of a lattice shell with the same span and no reinforcing plate. The magnitude of the increase for the lattice shell model with no defects is higher than that of the model with defects.

(2) The NSCLS is a defect-sensitive structure. The influence of geometric defects on its stable bearing capacity is very obvious. The results of comparing the lattice shell models show that the sensitivities of the two types of structures are different. The bearing capacity of the defective model with no plate is approximately 7% lower than that of the model without defects. In the combined lattice shell model, the bearing capacity of the model with defects is 12% lower than that of the model without defects.

(3) The finite element analysis results of applying the random defect mode method show that the theoretical failure patterns of the experimental models are basically consistent with those that were measured in the tests. The average difference between the theoretical stable bearing capacity and the experimental value is 5.7%. The theoretical load-displacement curves are also very close to the ones that were obtained in the tests. This indicates that the random defect mode method with a truncated Gaussian distribution is reasonable and reliable. It has sufficient accuracy. It can be used in structural analysis and design in practical engineering.

(4) As the lowest-order buckling mode is unable to characterize the most unfavorable distribution of defects in the new structure, the critical load obtained using the uniform defect mode is frequently not the smallest critical load. Therefore, the existing specification of defect sizes can no longer be applicable. By calculating and analyzing nearly 20,000 NSCLS, we find that after the initial geometric defect value of the combined lattice shell reaches 3/1000 of the shell's span, the stable bearing capacity decreases sharply. We recommend that the values be used as the maximum defect size for the combined lattice shell. The studies in this paper provide a theoretical basis for future design specifications for new composite lattice shell structures.

Acknowledgments: This study was supported by the Natural Science Foundation of China under Grant No. 51578136.

Author Contributions: Caiqi Zhao and Yangjian Zhao conceived and designed the experiments; Yangjian Zhao performed the experiments; Jun Ma and Yangjian Zhao analyzed the data; Jun Ma and Caiqi Zhao contributed reagents/materials/analysis tools; Caiqi Zhao wrote the paper.

Conflicts of Interest: The authors declare no conflict of interest.

References

1. Ma, Y.X.; Zheng, Y.D.; Meng, H.Y.; Song, W.H.; Yao, X.F.; Lv, H.X. Heterogeneous PVA hydrogels with micro-cells of both positive and negative Poisson's ratios. *J. Mech. Behav. Biomed. Mater.* **2013**, *23*, 22–31. [CrossRef] [PubMed]

2. Dirks, J.-H.; Dürr, V. Biomechanics of the stick insect antenna: Damping properties and structural correlates of the cuticle. *J. Mech. Behav. Biomed. Mater.* **2011**, *4*, 2031–2042. [CrossRef] [PubMed]

3. Koester, K.J.; Barth, H.D.; Ritchie, R.O. Effect of aging on the transverse toughness of human cortical bone: Evaluation by R-curves. *J. Mech. Behav. Biomed. Mater.* **2011**, *4*, 1504–1513. [CrossRef] [PubMed]

4. Donius, A.E.; Liu, A.; Berglund, L.A.; Ulrike, G.K. Wegst Superior mechanical performance of highly porous, anisotropic nanocellulose–montmorillonite aerogels prepared by freeze casting. *J. Mech. Behav. Biomed. Mater.* **2014**, *37*, 88–99. [CrossRef] [PubMed]

5. Tuo, W.Y.; Chen, J.X.; Wu, Z.S.; Xie, J.; Wang, Y. Characteristics of the tensile mechanical properties of fresh and dry forewings of beetles. *Mater. Sci. Eng. C* **2016**, *65*, 51–58. [CrossRef] [PubMed]

6. Chen, J.X.; Zu, Q.; Wu, G.; Xie, J. Review of beetle forewing structure and biomimetic applications in China(II). *Mater. Sci. Eng. C* **2015**, *50*, 620–633. [CrossRef] [PubMed]
7. Chen, J.X.; Gu, C.; Guo, S.; Wan, C.; Wang, X.; Xie, J.; Hu, X. Integrated honeycomb technology motivated by the structure of beetle forewings. *Mater. Sci. Eng. C* **2012**, *32*, 1813–1817. [CrossRef]
8. Zhang, X.M.; Liu, C.; Chen, J.X.; Tao, Y.; Gu, Y. The influence mechanism of processing holes on the flexural properties of biomimetic integrated honeycomb plates. *Mater. Sci. Eng. C* **2016**, *69*, 798–803. [CrossRef] [PubMed]
9. Chen, J.X.; He, C.L.; Gu, C.L.; Liu, J.X.; Mi, C.W.; Guo, J.S. Compressive and flexural properties biomimetic integrated honeycomb plates. *Mater. Des.* **2014**, *64*, 214–220. [CrossRef]
10. Chen, J.X.; Tuo, W.Y.; Zhang, X.M.; He, C.L.; Xie, J.; Liu, C. Compressive failure modes and parameter optimization of the trabecular structure of biomimetic fully integrated honeycomb plates. *Mater. Sci. Eng. C* **2016**, *69*, 255–261. [CrossRef] [PubMed]
11. Bourada, M.; Tounsi, A.; Houari, M.S.A.; Adda Bedia, E.A. A new four-variable refined plate theory for thermal buckling analysis of functionally graded sandwich plates. *J. Sand. Struct. Mater.* **2011**, *14*, 5–33. [CrossRef]
12. Zyniszewski, S.S.; Smith, B.H.; Hajjar, J.F.; Arwade, S.R.; Schafer, B.W. Local buckling strength of steel foam sandwich panels. *Thin-Walled Struct.* **2012**, *59*, 11–19. [CrossRef]
13. Katunin, A. Vibration-based spatial damage identification in honeycomb-core sandwich composite structures using wavelet analysis. *J. Compos. Struct.* **2014**, *118*, 385–391. [CrossRef]
14. Zhao, C.Q.; Ma, J. Assembled Honeycombed Sheet Light Empty Stomach Building and Roof Structure System. Patent Application No. 200810100745.X, 21 April 2010.
15. Zhao, C.Q.; Ma, J.; Tao, J. Experimental study on load capacity of new fabricated honeycomb panel open-web roof structures. *J. Southeast Univ. (Nat. Sci. Ed.)* **2014**, *44*, 626–630.
16. Zhao, C.Q.; Zheng, W.D.; Ma, J.; Zhao, Y.J. Lateral compressive buckling performance of aluminum honeycomb panels for long-span hollow core roofs. *Materials* **2016**, *9*, 444. [CrossRef] [PubMed]
17. Zhao, C.Q.; Zheng, W.D.; Ma, J.; Zhao, Y.J. Shear strength of different bolt connectors on large span aluminium alloy honeycomb sandwich structure. *Appl. Sci.* **2017**, *7*, 450. [CrossRef]
18. Liu, H.B.; Ding, Y.Z.; Chen, Z.H. Static stability behavior of aluminum alloy single-layer spherical latticed shell structure with Temcor joints. *Thin-Walled Struct.* **2017**, *46*, 82–89. [CrossRef]
19. Sugizaki, K.; Kohmura, S.; Hangai, Y. Experimental study on structural behaviour of an aluminum single-layer lattice shell. *Trans. AIJ* **1996**, *61*, 113–122. [CrossRef]
20. Hiyama, Y.; Takashima, H.; Iijima, T. Experiments and analyses of unit single-layer reticular domes using aluminum ball joints for the connections. *Trans. AIJ* **1999**, *64*, 33–40. [CrossRef]
21. Guo, H.N.; Xiong, Z.; Luo, Y.F. Block tearing and local buckling of aluminum alloy gusset joint plates. *KSCE J. Civ. Eng.* **2016**, *20*, 820–831. [CrossRef]
22. Zhang, Z.C.; Wang, X.D. Structural design and analysis of aluminum dome for Caofeidian coal storage. *Key Eng. Mater.* **2016**, *710*, 396–401. [CrossRef]
23. Gui, G.Q.; Wang, Y.E. Nonlinear stability analysis of single-layer aluminum alloy reticulated spherical shells. *Eng. Mech.* **2006**, *23*, 32–35.
24. Attard, M.M.; Hunt, G.W. Sandwich Column Buckling: A Hyperelastic Formulation. *J. Solids Struct.* **2008**, *45*, 5540–5555. [CrossRef]
25. Boudjemai, A.; Amri, R.; Mankour, A.; Salem, H.; Bouanane, M.H.; Boutchicha, D. Modal analysis and testing of hexagonal honeycomb plates used for satellite structural design. *Mater. Des.* **2012**, *35*, 266–275. [CrossRef]
26. Ravishankar, B.; Sankar, B.V.; Haftka, R.T. Homogenization of integrated thermal protection system with rigid insulation bars. In Proceedings of the 51st AIAA/ASCE/AHS/ASC Structures, Structural Dynamics and Materials Conference, Orlando, FL, USA, 12–15 April 2010; American Institute of Aeronautics and Astronautics: Boston, MA, USA, 2010.
27. Standardization Administration of China JGJ7-2010. *Technical Specification for Spatial Lattice Structures*; Standard Press of China: Beijing, China, 2010.

applied sciences

MDPI

Article

Manufacturing of Non-Stick Molds from Pre-Painted Aluminum Sheets via Single Point Incremental Forming

Oscar Rodriguez-Alabanda, Miguel A. Narvaez, Guillermo Guerrero-Vaca and Pablo E. Romero *

Department of Mechanical Engineering, University of Cordoba, Medina Azahara Avenue 5, 14071 Cordoba, Spain; orodriguez@uco.es (O.R.-A.); miguelnarvaez1991@gmail.com (M.A.N.); guillermo.guerrero@uco.es (G.G.-V.)
* Correspondence: p62rocap@uco.es; Tel.: +34-957-212-235

Received: 29 May 2018; Accepted: 18 June 2018; Published: 20 June 2018

Featured Application: In this work we propose a method for deforming a pre-coated metal sheet via SPIF. The influence of the pitch and feed-rate on the dimensional precision has been studied. This work is of interest for companies dedicated to the manufacture of molds for the agri-food sector, which can directly manufacture prototypes or short series by this system.

Abstract: The process of single point incremental formation (SPIF) awakens interest in the industry of mold manufacturing for the food industry. By means of SPIF, it is possible to generate short series of molds or mold prototypes at low cost. However, these industries require such molds to be functional (non-sticky) and to have an adequate geometry accuracy. This study presents a technique that enables direct manufacturing of molds from pre-coated sheets with non-stick resins. It has also studied the influence of two technological variables in the process (feed-rate and pitch) for different geometrical parameters of the mold. Low values of these variables result in a lower overall error in the profile obtained. However, in order to obtain greater detail in particular parameters (angles, depth), it is necessary to use higher values of feed-rate and pitch.

Keywords: SPIF; non-stick coatings; pre-painted metal sheet

---

1. Introduction

Single point incremental formation (SPIF), patented by Leszak in 1967 [1] and later developed by Prof. Matsubara [2] and Kitazawa et al. [3], consists of the progressive deformation of a metal sheet by means of a steel punch at the top of a machining center or robot. The forming is done in such a way that neither presses nor dies are necessary [4].

The SPIF technique enables the manufacturing of prototypes from sheet materials. By means of SPIF, forming can be done to all types of metals and alloys: aluminum, steel, stainless steel, tin, copper, titanium, among others [5,6]. Polymers can also be worked with [7]: polyamide (PA), polyethylene (PE), polyvinyl chloride (PVC), polystyrene (PS), polypropylene (PP), polycarbonate (PC), polyethylene terephthalate (PET). The advantages and disadvantages of the process are summarized in Table 1.

As well as prototypes, with SPIF, it is also possible to obtain short manufacturing series at low cost. Petek et al. [8] determined that SPIF technology is more profitable than deep drawing up to a batch size of 600 pieces. Ingarao et al. [9] affirm that SPIF is a more efficient process than deep forming processes which produce cuttings and scrap metal.

Table 1. Advantages and disadvantages of single point incremental formation (SPIF) [10].

Advantages	Disadvantages
It can be done in a machining center	The equipment must be managed by experienced and qualified operators
The changes in design may be done easily and quickly	The processing time is longer
The strengths within the material are relatively low	Elastic recuperation is produced
The pieces are produced directly from a electronic file	The process is limited to medium to small batches
The dimensions of the pieces are only restricted by the dimensions of the tool machinery	The forming of angles of 90° tends to be limited

The process has evolved in recent years [11], and its use has increased in different sectors, such as automotive [12–16], medicine [17–19], energy [20], architecture [21], and art [22]. One of the least explored uses of this process is in the manufacturing of molds. Fiorotto et al. [23] have proposed the use of molds made by means of SPIF directed at compound material sheeting. Appermont et al. [24] and Afonso et al. [25] have considered its use in thermoforming operations. Recently, it has been tested in rotomolding processes, and in other polymer processing techniques [26].

In most cases, for the demolding process, it is desirable for the mold to have a non-stick coating. These coatings may be of several types, although the most common are those made with resins that are rich in fluoropolymers [27]. Such resins may be applied to the metal sheets before forming, so that the manufactured piece is the final one. However, until now the friction between the tool and the sheet complicated the forming process with pre-coated SPIF sheeting [28].

In order to avoid direct contact between the tool and the coating, in this study, the dummy technique [29,30] has been used: a dummy sheet is placed over the sheet to be formed, and the punch simultaneously forms both sheets, without any direct contact with the lower pre-coated sheet. This novel method enables the application of the SPIF process to sheets pre-coated in resins rich in fluoropolymers, and thus, among other applications, directly manufacture prototypes or short series of non-stick molds.

The aluminum–magnesium alloy EN AW 5754 H32 is commonly used to manufacture molds in the food industry, as it comes into contact with food and has good formability [31]. The incremental forming of this material has been studied previously by Verbert [32], Behera [33], and Gupta & Jeswiet [34].

For the industrial use of SPIF, it is imperative to reduce springback and achieve good geometrical accuracy [10]. There are previous studies that have analyzed this matter with aluminum alloys. Ambrogio et al. [35,36] focused on the improvement of the dimensional accuracy of truncated cone pieces from sheets of EN AW 1050-O. Guzman et al. [37] advanced in the accuracy reached using SPIF in two-slope cones manufactured in EN AW 3003. However, no previous studies have been found in the literature concerning the geometrical accuracy reached using SPIF with sheets of EN AW 5754 H32, nor authors who have approached this subject using the dummy method.

This work has two main aims. The first is to present and describe the use of the dummy method in incremental forming for the direct manufacturing of prototypes or short series of molds. The second aim is to study the influence of two technological parameters, pitch and feed-rate, in the accuracy reached in the manufacturing of truncated pyramidal geometry. The piece was manufactured using SPIF in EN AW 5754 H32 aluminum–magnesium sheets, pre-coated with resins rich in polytetrafluoroethylene (PTFE). The geometrical variables measured were initial and final angle, minimum thickness, depth, height, normalized springback, and area variation between theoretical and experimental profile.

The structure of the study is the following: Section 2 describes the equipment and material used in both the forming process and the measuring; Section 3 shows the results, which are discussed in Section 4; finally, the conclusions are presented in Section 5.

2. Materials and Methods

The dummy method in SPIF consists of the simultaneous deforming of two overlapping sheets [38]. This method has been used to improve the surface finish [29] and to deform previously welded sheets by means of friction stir welding [30]. In this case, the dummy method was used to manufacture non-stick molds from sheets pre-coated with PTFE resin (Figure 1).

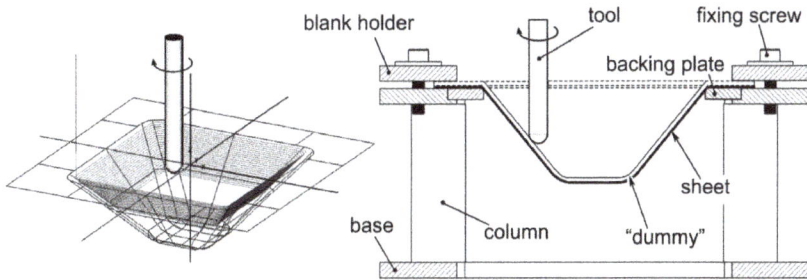

Figure 1. Schematic representation of SPIF process.

The molds were made with aluminum–magnesium EN AW 5754-H32 sheets, the dimensions being 210 × 210 mm and 1.2 mm thick. On one side, a circular pattern was printed to control the deforming undergone at each point [39]. The other side was covered with a resin-rich PTFE. To protect this re-covering during the incremental deforming, a PVC dummy sheet of 210 × 210 mm and 3.0 mm thick was used.

The design chosen for the molds was a truncated pyramidal geometrical shape, as it is of industrial interest and facilitates the geometrical characteristics. Several tests were carried out in order to determine the maximum depth and angle that the sheets were able to reach. The final dimensions are shown in Figure 2.

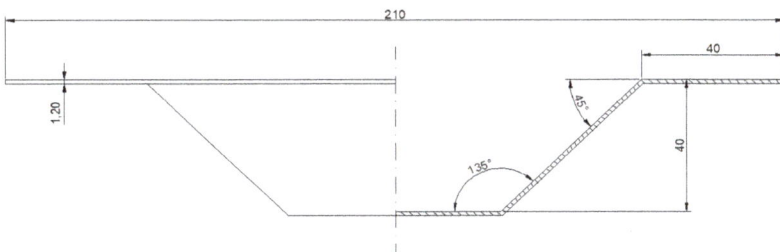

Figure 2. Dimensions of the mold used in the tests (unit: mm).

The pieces were drawn using SolidWorks parametric software. Once drawn, the models were exported to the MasterCAM program, where the strategy was defined, and the technological parameters of the process were determined. This being done, the MasterCAM automatically generated the NC programs. The tool path strategy used was contour-parallel.

Nine tests were carried out, with different pitch and feed-rate values. For the pitch, the values 0.3, 0.6 and 1.0 mm were used; for the feed-rate, the values were 1800, 2000 and 2200 mm/min. The spindle speed in all cases was set at 750 RPM.

The geometrical parameters and technologies were selected according to database collected from the literature (Table 2) and after carrying out several sensitivity tests (Table 3). Breakage of any of the materials was avoided (PVC, PTFE, EN-AW 5754), since the aim was to attain healthy molds that could be studied geometrically.

Table 2. Parameters used in literature for SPIF of PVC and EN-AW 5754.

Authors	Material	Feed-Rate (mm/min)	Spindle Speed (RPM)	Pitch (mm)
Franzen et al. [40]	PVC	1500	Free	0.5
Silva, Alves & Martins [41]	PVC	1000	Free	0.5
Martins et al. [7]	PVC	1500	Free	0.5
Zhang, Wang & Zhang [42]	PVC	2000	2000	0.5/1.0
Medina-Sanchez et al. [43]	PVC	1500	500	0.5
Ambrogio et al. [44]	EN-AW 5754	2000/8000	2500	0.5
Aerens et al. [45]	EN-AW 5754	2000	–	0.5
Ingarao et al. [9]	EN-AW 5754	2000	200	1.0
Gupta & Jeswiet [34]	EN-AW 5754	3000/7500	1000/2000	0.2

Table 3. Results of preliminary tests.

Test	Angle (°)	Depth (mm)	Pitch (mm)	Spindle Speed (RPM)	Feed-Rate (mm/min)	Results
#1	50	60	1.2	750	2200	PVC sheet breaks
#2	45	65	1.2	750	2200	PVC sheet breaks
#3	45	60	1.2	750	2200	OK
#4	55	50	1.2	750	2200	Al sheet break
#5	50	50	1.2	750	2200	OK

It is important to bear in mind that during the process, there is a simultaneous deformation of two sheets with very different mechanical properties and behaviors. The dummy sheet (PVC) is that which is in contact with the punch, so that, although the intention is to deform the aluminum, it is necessary to work with values that are compatible with the PVC (Table 2).

The tests were carried out at a model QP2026-L Chevalier machining center, equipped with a Fanuc numerical control. The tool used for the forming was a 14.68 mm diameter punch, made up of two pieces: an EN AW 2024 aluminum body and a stainless steel spherical tip. SAE 30 mineral oil was used for lubrication, following the recommendations of Azevedo et al. [46]. Experiment set-up is shown in Figure 3.

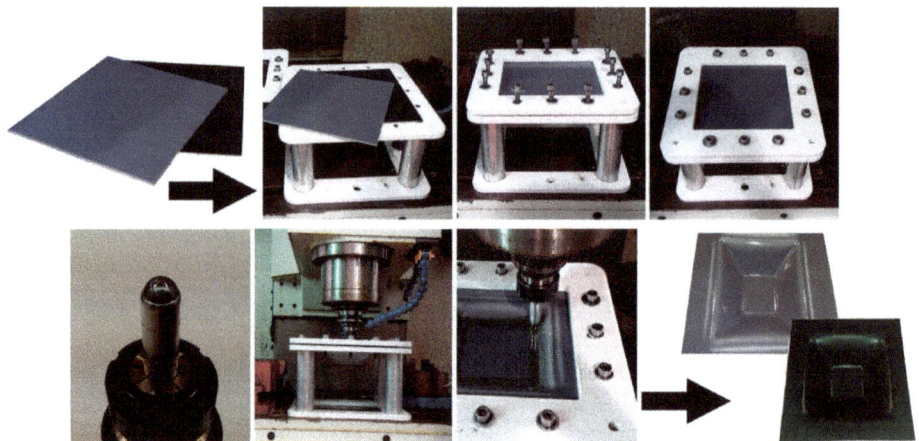

Figure 3. Experiment set-up and manufactured pieces.

Once the pieces were manufactured, different geometrical parameters were measured (Figure 4): minimum thickness, depth, height, initial angle, elastic recovery, and area of deviation between theoretical and real profile.

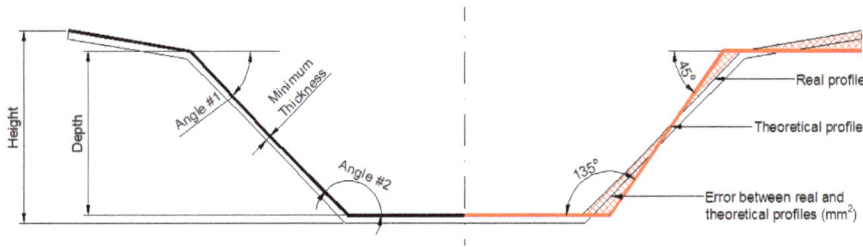

Figure 4. Geometrical characterization of the part and error respect to theoretical profile.

Depth was measured using a depth gauge and gauge blocks. Four measurements were taken in each mold. Height was measured on marble, using a height measure and four different measurements were also taken in each mold.

Angle #1 was measured with a goniometer. From this angle measured (actual) and the theoretical angle (CAD), the normalized springback was calculated, by means of the expression Equation (1), proposed by Martins et al. [7].

$$S_n = \frac{(\Psi_{CAD} - \Psi_{actual})}{\Psi_{CAD}} \times 100 \tag{1}$$

To measure the thickness, the molds were cut in half using a belt saw. Once this was done, different measurements were taken along the profile (one measurement every 10 mm) using a 0.01 precision micrometer. Fifty-six measurements were taken for each piece.

Once cut, the pieces were photographed. The images were imported from the AutoCAD program. Once scaled, the sections were drawn over them, which were used to measure angle #2. Besides this, the sections were compared with the theoretical profile of the mold. The area enclosed between them gives the overall error existing between both profiles.

3. Results

Figure 5 shows the exterior angle or angle #1 measured in each of the molds manufactured via SPIF with a dummy sheet. The theoretical angle sought after was 45°. Figure 6 shows the interior angle or angle #2. In this case, the theoretical angle sought after was 135°. The minimum thickness measured in the sections of the different pieces is shown in Figure 7. As can be observed, it shows no great variations for the different cases studied.

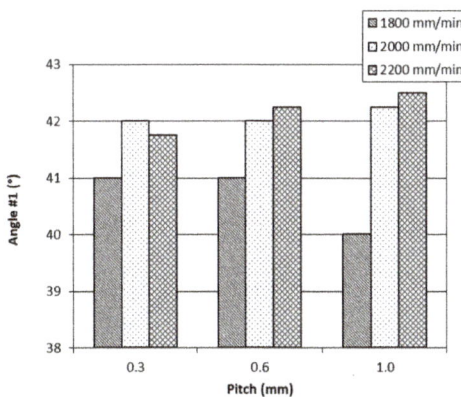

Figure 5. Experimental angle #1 obtained for different pitches and feed-rates.

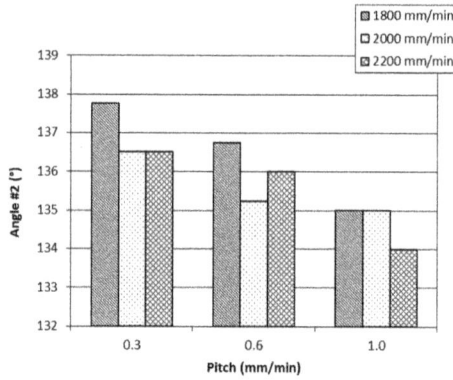

Figure 6. Experimental angle #2 obtained for different pitches and feed-rates.

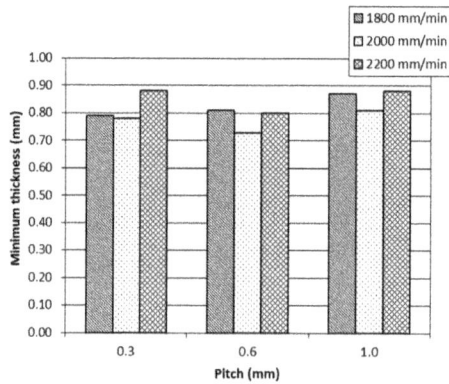

Figure 7. Minimum thickness obtained in the different tests.

Figure 8 shows the average height and depth values measured in each mold. As can be appreciated, both parameters are closer to the expected theoretical measurements for greater values of pitch and feed-rate.

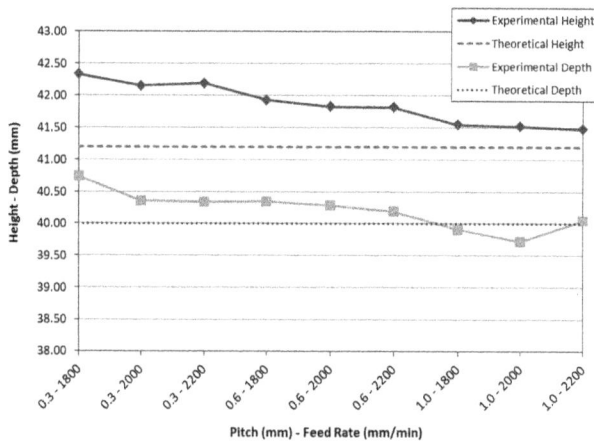

Figure 8. Height and depth of the molds, for different pitches and feed-rates.

The normalized springback measured in the different tests can be seen in Figure 9. Here, the pitch is not seen to be an influencing factor. However, it appears that lower values of springback are associated with high values of feed-rate speed.

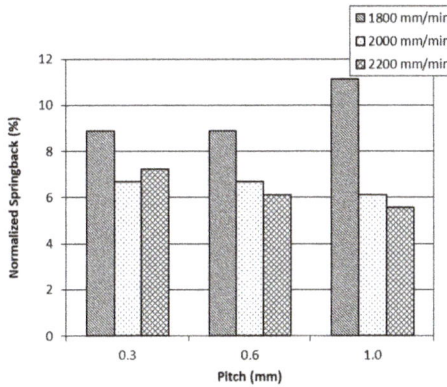

Figure 9. Normalized springback, for different pitches and feed-rates.

The area between the theoretical profile and the real one for each mold is shown in Figure 10. This area gives an idea of the overall error existing between the sought-after piece and the one obtained.

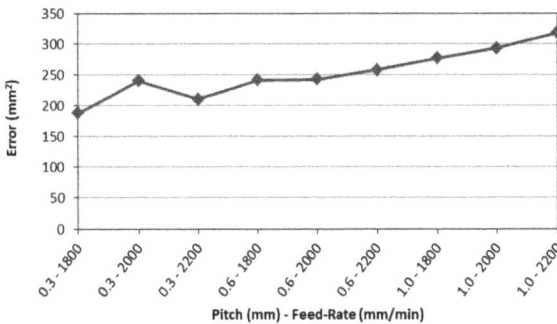

Figure 10. Area between real and theoretical profiles (global error).

Figure 11 shows the forming limit diagram (FLD) corresponding to the manufactured molds with a feed-rate equal to 2000 mm/min. As can be observed, the greater the pitch, the greater the main deformations.

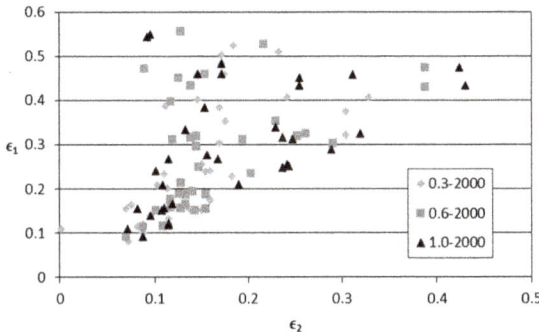

Figure 11. Format limit diagram for molds (2000 mm/min).

4. Discussion

The dummy method was inserted into the SPIF by Skjoedt et al. [38]. The initial aim was to improve the surface finish of the manufactured metallic pieces by incremental deforming [29]. Later, the dummy method was used to deform welded sheets by means of friction stir welding [30]. In the present study, the dummy method has been used to conform, directly via SPIF, pre-coated molds rich in a PTFE resin.

As well as presenting this novel method for directly manufacturing non-stick molds by incremental deforming, the present work has also studied the influence of both technological parameters (pitch and feed-rate) on the dimensional precision reached during the incremental forming of molds from pre-coated sheets using the dummy technique.

After analyzing the results, we can affirm the following:

- The minimum thickness appears to show no relation to any of the technological parameters studied (Figure 7). This result is coherent with the law of sine [47]. According to this law, the theoretical thickness can be calculated as a product of the initial thickness (1.20 mm) and the sine of the angle (45°). The theoretical value calculated (0.84 mm) coincides with the value measured experimentally.
- The differences between the theoretical profile and the real profile of the mold are minor, in general terms, when using low values of pitch and feed-rate (Figure 10). These results coincide with those obtained by Hussain, Lin & Hayat [48] (greater pitches provoke greater deviations), Maqbool & Bambach [49] (lower pitches provoke lower geometrical accuracy) and Radu & Cristea [50] (high values of the feed-rate lead to a low-dimensional accuracy).
- However, low values of pitch and feed-rate result in certain geometrical features (depth, height, angle #1, normalized springback, angle #2) moving away from the sought-after theoretical values:

  - The depth and height approach the theoretical ones with high pitch and feed-rate values (Figure 8).
  - Angle #1 approaches the sought-after value (45°) when high feed-rate values are programmed (Figure 5). Logically, the normalized springback shows the same behavior (Figure 9).
  - Angle #2 (135°) approaches the sought-after value when greater pitch values are used (Figure 6).

- An increase in pitch increases the deformability (Figure 11). This result coincides with that presented by Liu et al. [51].

As can be appreciated, the test carried out with a greater pitch and feed-rate attains better results for the particular variables (height, depth, initial angle, maximum angle, minimum thickness). At the other end, the test carried out with a lower pitch and feed-rate manages to minimize the deviation area between the profile obtained and the expected theoretical one.

To sum up, it can be affirmed that the results obtained relative to geometrical accuracy in the dummy method are similar to those obtained in the incremental deforming processes of a single sheet. To solve this problem, Bambach, Araghi & Hirt [13] proposed a multistage strategy to improve geometrical accuracy. Following this idea, the authors are going to try a multi-stage strategy: firstly, high values of pitch and feed-rate will be used to get a good reproduction of the geometry details; secondly, lower values of pitch and feed-rate will be used to provide a lesser deviation in the real compared to the theoretical profile.

5. Conclusions

The present study has analyzed the geometrical accuracy obtained in the SPIF process of sheets pre-coated with resins rich in PTFE, using the dummy method. To do so, truncated pyramidal shapes were formed in EN-AW 5754 sheets of aluminum–magnesium with a PVC dummy sheet, using different pitch and feed-rate values.

The experiments carried out show that there is no apparent relation between some of the geometrical parameters studied (minimum thickness) and the technological parameters cited. However, it has been proven that there is a relation between the remaining parameters: the normalized springback value decreases with greater feed-rate values; height and depth get closer to the sought-after values when greater values of pitch and feed-rate are programmed; the area between the real and the theoretical profile (a measurement of error) also increase when using greater pitch and feed-rate values.

These results are of great interest to the local industrial sector involved in the manufacture of molds and trays for food products which, in the SPIF with dummy process, find a method for manufacturing prototypes from pre-coated sheets.

Future studies will address the geometrical accuracy obtained in incremental deforming processes using multi-stage strategy. In the initial stages, high values of pitch and feed-rate will be used in order to guarantee a good reproduction of the geometry details. In the final stages, lower values of pitch and feed-rate will be used that will provide a lesser deviation in the real compared to the theoretical profile. Besides this, the exact angles and maximum depths that can be reached by sheets of this material with different thicknesses will be studied, using different values of spindle speed and different tool diameters.

Author Contributions: O.R.-A. and P.E.R. conceived and designed the experiments; O.R.-A. and M.A.N. performed the experiments; M.A.N., G.G.-V. and P.E.R. analyzed the data; O.R.-A. and P.E.R. wrote the paper; G.G.-V. revised the paper.

Funding: This research received no external funding.

Conflicts of Interest: The authors declare no conflict of interest.

References

1. Leszak, E. Apparatus and Processs for Incremental Dieless Forming. U.S. Patent 3342051, 19 September 1967.
2. Matsubara, S. Incremental backward bulge forming of a sheet metal with a hemispherical head tool. *Jpn. Soc. Technol. Plast.* **1994**, *35*, 1311–1316.
3. Kitazawa, K.; Wakabayashi, A.; Murata, K.; Yaejima, K. Metal-flow phenomena in computerized numerically controlled incremental stretch-expanding of aluminum sheets. *J. Jpn. Inst. Light Met.* **1996**, *46*, 65–70. [CrossRef]
4. Jeswiet, J.; Micari, F.; Hirt, G.; Bramley, A.; Duflou, J.; Allwood, J. Asymmetric Single Point Incremental Forming of Sheet Metal. *CIRP Ann. Manuf. Technol.* **2005**, *54*, 88–114. [CrossRef]
5. Afonso, D.; de Sousa, R.; Torcato, R. Integration of design rules and process modelling within SPIF technology—A review on the industrial dissemination of single point incremental forming. *Int. J. Adv. Manuf. Technol.* **2018**, *94*, 4387–4399. [CrossRef]
6. McAnulty, T.; Jeswiet, J.; Doolan, M. Formability in single point incremental forming: A comparative analysis of the state of the art. *CIRP J. Manuf. Sci. Technol.* **2017**, *16*, 43–54. [CrossRef]
7. Martins, P.A.F.; Kwiatkowski, L.; Franzen, V.; Tekkaya, A.E.; Kleiner, M. Single point incremental forming of polymers. *CIRP Ann. Manuf. Technol.* **2009**, *58*, 229–232. [CrossRef]
8. Petek, A.; Gantar, G.; Pepelnjak, T.; Kuzman, K. Economical and Ecological Aspects of Single Point Incremental Forming Versus Deep Drawing Technology. *Key Eng. Mater.* **2007**, *344*, 931–938. [CrossRef]
9. Ingarao, G.; Ambrogio, G.; Gagliardi, F.; Di Lorenzo, R. A sustainability point of view on sheet metal forming operations: Material wasting and energy consumption in incremental forming and stamping processes. *J. Clean. Prod.* **2012**, *29–30*, 255–268. [CrossRef]
10. Echrif, S.B.M.; Hrairi, M. Research and Progress in Incremental Sheet Forming Processes. *Mater. Manuf. Process.* **2011**, *26*, 1404–1414. [CrossRef]
11. Behera, A.K.; de Sousa, R.A.; Ingarao, G.; Oleksik, V. Single point incremental forming: An assessment of the progress and technology trends from 2005 to 2015. *J. Manuf. Process.* **2017**, *27*, 37–62. [CrossRef]
12. Zha, G.; Xu, J.; Shi, X.; Zhou, X.; Lu, C. Forming process of automotive body panel based on incremental forming technology. *Metall. Min. Ind.* **2015**, *12*, 350–357.
13. Bambach, M.; Taleb Araghi, B.; Hirt, G. Strategies to improve the geometric accuracy in asymmetric single point incremental forming. *Prod. Eng.* **2009**, *3*, 145–156. [CrossRef]

14. Governale, A.; Lo Franco, A.; Panzeca, A.; Fratini, L.; Micari, F. Incremental Forming Process for the Accomplishment of Automotive Details. *Key Eng. Mater.* **2007**, *344*, 559–566. [CrossRef]
15. Romero, P.E.; Aguilar-Contreras, F.J.; Dorado, R.; Lopez-Garcia, R. Rapid prototyping for automotive industry via incremental sheet forming. *DYNA* **2013**, *88*, 581–590.
16. Amino, M.; Mizoguchi, M.; Terauchi, Y.; Maki, T. Current status of "Dieless" Amino's incremental forming. *Procedia Eng.* **2014**, *81*, 54–62. [CrossRef]
17. Ambrogio, G.; De Napoli, L.; Filice, L.; Gagliardi, F.; Muzzupappa, M. Application of Incremental Forming process for high customised medical product manufacturing. *J. Mater. Process. Technol.* **2005**, *162–163*, 156–162. [CrossRef]
18. Castelan, J.; Schaeffer, L.; Daleffe, A.; Fritzen, D.; Salvaro, V.; Da Silva, F.P. Manufacture of custom-made cranial implants from DICOM® images using 3D printing, CAD/CAM technology and incremental sheet forming. *Rev. Bras. Eng. Biomed.* **2014**, *30*, 265–273. [CrossRef]
19. Centeno, G.; Bagudanch, I.; Morales-Palma, D.; García-Romeu, M.L.; Gonzalez-Perez-Somarriba, B.; Martinez-Donaire, A.J.; Gonzalez-Perez, L.M.; Vallellano, C. Recent Approaches for the Manufacturing of Polymeric Cranial Prostheses by Incremental Sheet Forming. *Procedia Eng.* **2017**, *183*, 180–187. [CrossRef]
20. Duflou, J.R.; Lauwers, B.; Verbert, J. Study on the achievable accuracy in single point incremental forming. In *Advanced Methods in Material Forming*; Springer: Berlin/Heidelberg, Germany, 2007; pp. 251–262.
21. Afonso, D.; Alves de Sousa, R.; Torcato, R.; Sousa, J.P.; Santos, R.; Valente, R. Case studies on industrial applicability of single point incremental forming. In Proceedings of the Ciência 2017—Encontro com Ciência e Tecnologia en Portugal, Lisbon, Portugal, 3–5 July 2017.
22. Kalo, A. N-Bowls. Available online: www.ammarkalo.com (accessed on 4 April 2018).
23. Fiorotto, M.; Sorgente, M.; Lucchetta, G. Preliminary studies on single point incremental forming for composite materials. *Int. J. Mater. Form.* **2010**, *3*, 951–954. [CrossRef]
24. Appermont, R.; Van Mieghem, B.; Van Bael, A.; Bens, J.; Ivens, J.; Vanhove, H.; Behera, A.K.; Duflou, J. Sheet-metal based molds for low-pressure processing of thermoplastics. *PMI* **2008**, 383–388.
25. Afonso, D.; De Sousa, R.A.; Torcato, R. Testing single point incremental forming molds for thermoforming operations. *AIP Conf. Proc.* **2016**, *1769*, 060016. [CrossRef]
26. Afonso, D.; Pires, L.; de Sousa, R.A.; Torcato, R. Direct rapid tooling for polymer processing using sheet metal tools. *Procedia Manuf.* **2017**, *13*, 102–108. [CrossRef]
27. Ruiz-Cabello, F.J.M.; Rodríguez-Criado, J.C.; Cabrerizo-Vílchez, M.; Rodríguez-Valverde, M.A.; Guerrero-Vacas, G. Towards super-nonstick aluminized steel surfaces. *Prog. Org. Coat.* **2017**, *109*, 135–143. [CrossRef]
28. Katajarinne, T.; Vihtonen, L.; Kivivuori, S. Incremental forming of colour-coated sheets. *Int. J. Mater. Form.* **2008**, *1*, 1175–1178. [CrossRef]
29. Skjoedt, M.; Silva, M.B.; Bay, N.; Martins, P.A.F. Single point incremental forming using a dummy sheet. In Proceedings of the 2nd International Conference on New Forming Technologies, Bremen, Germany, 20–21 September 2007; pp. 267–276.
30. Silva, M.B.; Skjoedt, M.; Vilaça, P.; Bay, N.; Martins, P.A.F. Single point incremental forming of tailored blanks produced by friction stir welding. *J. Mater. Process. Technol.* **2009**, *209*, 811–820. [CrossRef]
31. Guerrero-Vacas, G. *Comparative Analysis of the Removal Processes of Fluoropolymer Anti-Adherent Coatings on Metallic Surfaces between Laser and Pyrolytic Technologies*; University of Malaga: Malaga, Spain, 2013.
32. Verbert, J. *Computer Aided Process Planning for Rapid Prototyping with Incremental Sheet Forming Techniques*; Katholieke Universiteit Leuven: Leuven, Belgium, 2010.
33. Behera, A.K. *Shape Feature Taxonomy Development for Toolpath Optimization in Incremental Sheet Forming*; Katholieke Universiteit Leuven: Leuven, Belgium, 2013.
34. Gupta, P.; Jeswiet, J. Observations on Heat Generated in Single Point Incremental Forming. *Procedia Eng.* **2017**, *183*, 161–167. [CrossRef]
35. Ambrogio, G.; Costantino, I.; De Napoli, L.; Filice, L.; Fratini, L.; Muzzupappa, M. Influence of some relevant process parameters on the dimensional accuracy in incremental forming: A numerical and experimental investigation. *J. Mater. Process. Technol.* **2004**, *153–154*, 501–507. [CrossRef]
36. Ambrogio, G.; Cozza, V.; Filice, L.; Micari, F. An analytical model for improving precision in single point incremental forming. *J. Mater. Process. Technol.* **2007**, *191*, 92–95. [CrossRef]

37. Guzmán, C.F.; Gu, J.; Duflou, J.; Vanhove, H.; Flores, P.; Habraken, A.M. Study of the geometrical inaccuracy on a SPIF two-slope pyramid by finite element simulations. *Int. J. Solids Struct.* **2012**, *49*, 3594–3604. [CrossRef]

38. Skjoedt, M. Rapid Prototyping by Single Point Incremental Forming of Sheet Metal. Ph.D. Thesis, Technical University of Denmark, Kgs. Lyngby, Denmark, 2008.

39. Jackson, K.; Allwood, J. The mechanics of incremental sheet forming. *J. Mater. Process. Technol.* **2009**, *209*, 1158–1174. [CrossRef]

40. Franzen, V.; Kwiatkowski, L.; Martins, P.A.F.; Tekkaya, A.E. Single point incremental forming of PVC. *J. Mater. Process. Technol.* **2009**, *209*, 462–469. [CrossRef]

41. Silva, M.B.; Alves, L.M.; Martins, P.A.F. Single point incremental forming of PVC: Experimental findings and theoretical interpretation. *Eur. J. Mech. A/Solids* **2010**, *29*, 557–566. [CrossRef]

42. Zhang, X.; Wang, J.; Zhang, S. Study on Process Parameters on Single Point Incremental Forming of PVC. *Mater. Sci. Forum* **2017**, *878*, 74–80. [CrossRef]

43. Medina-Sánchez, G.; Torres-Jimenez, E.; Lopez-Garcia, R.; Dorado-Vicente, R.; Cazalla-Moral, R. Temperature influence on Single Point Incremental Forming of PVC parts. *Procedia Manuf.* **2017**, *13*, 335–342. [CrossRef]

44. Ambrogio, G.; Ingarao, G.; Gagliardia, F.; Di Lorenzo, R. Analysis of energy efficiency of different setups able to perform single point incremental forming (SPIF) processes. *Procedia CIRP* **2014**, *15*, 111–116. [CrossRef]

45. Aerens, R.; Duflou, J.R.; Eyckens, P.; van Bael, A. Advances in force modelling for SPIF. *Int. J. Mater. Form.* **2009**, *2*, 25–28. [CrossRef]

46. Azevedo, N.G.; Farias, J.S.; Bastos, R.P.; Teixeira, P.; Davim, J.P.; Alves de Sousa, R.J. Lubrication aspects during Single Point Incremental Forming for steel and aluminum materials. *Int. J. Precis. Eng. Manuf.* **2015**, *16*, 589–595. [CrossRef]

47. Jeswiet, J.; Hagan, E.; Szekeres, A. Forming parameters for incremental forming of aluminium alloy sheet metal. *Proc. Inst. Mech. Eng. Part B J. Eng. Manuf.* **2002**, *216*, 1367–1371. [CrossRef]

48. Hussain, G.; Gao, L.; Hayat, N. Forming parameters and forming defects in incremental forming of an aluminum sheet: Correlation, empirical modeling, and optimization: Part A. *Mater. Manuf. Process.* **2011**, *26*, 1546–1553. [CrossRef]

49. Maqbool, F.; Bambach, M. Dominant deformation mechanisms in single point incremental forming (SPIF) and their effect on geometrical accuracy. *Int. J. Mech. Sci.* **2018**, *136*, 279–292. [CrossRef]

50. Radu, M.C.; Cristea, I. Processing metal sheets by SPIF and analysis of parts quality. *Mater. Manuf. Process.* **2013**, *28*, 287–293. [CrossRef]

51. Liu, Z.; Li, Y.; Meehan, P.A. Experimental investigation of mechanical properties, formability and force measurement for AA7075-O aluminum alloy sheets formed by incremental forming. *Int. J. Precis. Eng. Manuf.* **2013**, *14*, 1891–1899. [CrossRef]

applied sciences

MDPI

Article

Stability of Cu-Precipitates in Al-Cu Alloys

Torsten E. M. Staab [1,*,†], Paola Folegati [2,†], Iris Wolfertz [3] and Martti J. Puska [4]

1 LCTM, Universität Würzburg, Röntgenring 11, D-97070 Würzburg, Germany
2 Politecnico Milano, Polo di Como, Via Anzani 42, I-22100 Como, Italy; paola.folegati@polimi.it
3 HISKP, Universität Bonn, Nußallee 14-16, D-53115 Bonn, Germany; iris.wolfertz@hiskp.uni-bonn.de
4 Department of Applied Physics, Aalto University, P.O. Box 11100, FI-00076 Aalto, Finland; martti.puska@aalto.fi
* Correspondence: Torsten.Staab@uni-wuerzburg.de; Tel.: +49-931-31-81460
† These authors contributed equally to this work.

Received: 30 May 2018; Accepted: 9 June 2018; Published: 20 June 2018

Abstract: We present first principle calculations on formation and binding energies for Cu and Zn as solute atoms forming small clusters up to nine atoms in Al-Cu and Al-Zn alloys. We employ a density-functional approach implemented using projector-augmented waves and plane wave expansions. We find that some structures, in which Cu atoms are closely packed on {100}-planes, turn out to be extraordinary stable. We compare the results with existing numerical or experimental data when possible. We find that Cu atoms precipitating in the form of two-dimensional platelets on {100}-planes in the fcc aluminum are more stable than three-dimensional structures consisting of the same number of Cu-atoms. The preference turns out to be opposite for Zn in Al. Both observations are in agreement with experimental observations.

Keywords: aluminum copper alloys; Guinier-Preston zones; precipitates; ab initio calculations; DFT-LDA

---

1. Introduction

Al-Cu alloys have been under active scientific research and technological development for more than 100 years because of their applications in light weight constructions [1,2]. Nowadays, they are especially important in aviation and automotive industry. Aluminum alloys show a rich variety of metastable and stable phases from which a few are ordered compounds. Since usually the surface energy is too large to form directly thermodynamically stable phases, alloying elements precipitate in a sequence of clusters, zones and metastable phases. Clusters are non-ordered, locally increased concentrations of solute atoms, zones are locally ordered but do not have a long-range ordering, while stable and metastable phases possess the latter [3]. Zones and some metastable phases are typically fully coherent with the matrix [3].

Precipitation and clustering phenomena of solute atoms in a light metal matrix are the reason for the superior properties of aluminum alloys, i.e., this results in a high strength at a small specific weight. The obtained mechanical properties arise from a suitable thermal treatment of these alloys [2]. Typically, after casting these age hardenable alloys are extruded or rolled to their final form. Thereafter they undergo a solution heat treatment at about $100 \ldots 150\,\text{K}$ below the melting point of aluminum in order to obtain the maximum solubility of the chosen alloying elements [3]. After the heat treatment the materials are quenched to room temperature to freeze-in the finely distributed solute atoms. Storing these alloys then at room temperatures causes the solute atoms to diffuse by the help of quenched-in vacancies [4]. Via this process the solute atoms form agglomerates, which grow subsequently in size. After storing Al-Cu alloys for some hours at room temperature the agglomerates become visible in X-ray diffraction patterns and they are called Guinier-Preston zones according to Guinier and Preston, who discovered them independently in 1938 [5–7]. Further storage at elevated temperature causes the growth of meta-stable Al-Cu phases: θ'' Al$_3$Cu and θ' Al$_2$Cu [3]. However, in new generation Al-Cu-Li alloys like AA2198, AA2050 or AA2199 besides Al-Cu-Li precipitates these Al-Cu-phases are detected as well [8,9]. Those Al-Cu-Li alloys are considered for the fuselage of new generation aircrafts due to their high strength and good welding behavior and have been, thus, subject to intense research in recent years (see e.g., [10,11]), while also Al-Cu alloys are still a matter of active research [12,13].

However, the understanding of the precipitation process in metallic alloys on the atomic level is still one of the main problems in materials science. It hampers a purposeful improvement of alloys, i.e., an alloy design as a bottom-up approach. Since the atomic structure of small, i.e., sub-nanometer, precipitates is difficult to access experimentally, numerical ab initio simulations are often the only way to obtain data on the geometry of atomic arrangements and their binding properties. Up to now, just a few numerical results on vacancy formation energies and di-vacancy binding energies in aluminum are available [14,15], which can be compared with accurate experimental data on vacancy formation energies in pure Al (see [16] for an overview). Only recently, research on vacancy binding with different isolated solute atoms has been published for Al [17,18] and Mg [19].

Results from ab initio calculations can be compared to experiments probing, e.g., vacancies by positron annihilation spectroscopy (PAS) [20–23] or individual elements by X-ray absorption [24,25] or small solute atom clusters employing the atom probe methods [26]. Moreover, the recently re-discovered X-ray absorption fine structure (XAFS) spectroscopy is sensitive for the atomic environment of, e.g., Cu-atoms [24,27]. For the two spectroscopic methods, PAS and XAFS, spectra can be calculated from first principles (For PAS see, e.g., ref. [28] and for XAFS [29]). The resulting spectra, which are related to trapping of positrons to vacancies or to the excitation of solute atoms like Cu and Zn by X-rays, depend strongly on the atomic positions around those defects. A comparison of the simulations with existing experimental data can be effectively used to search for an explanation of the clustering phenomena on the atomic level in different sample compositions and conditions. Thus, it can provide guidelines to metallurgists to perform thermal and mechanical treatments on Al-alloys in order to obtain the desired materials properties.

Specifically, the results of the present work clearly give an ab initio explanation, why in Al-Cu alloys copper precipitates on the {100}-planes, while for Al-Zn alloys three-dimensional (3D) agglomerates of Zn-atoms are formed. The reasons for these findings are easy to understand in the named simple two-component systems. However, his understanding will also pave the way for controlling processes taking place in actual technical alloys composed often of more than five elements.

The present paper is organized as follows. The computational schemes employed are presented in Section 2. Results on the formation energies of vacancies and di-vacancies are given in Section 3. Then we present vacancy-solute and solute-solute binding energies for clusters containing up to nine copper atoms. Section 4 contains a discussion – in particular, a comparison between the different employed calculation schemes is presented.

2. Methods: Computational Schemes

All our calculations are based on density functional theory (DFT) within the local density approximation (LDA). In some cases a comparison with the generalized gradient approximation (GGA) of DFT has been performed. The computations are carried out using the plane-wave code VASP [30,31], implemented with the projector augmented-wave (PAW) method to account for the valence electron-ion core interaction.

In our VASP calculations, we have employed supercells of different sizes—namely 64, 108, 128, 144 and 192 atoms per supercell are used to check the influence of finite size effects on the relaxation of the atoms and the derived total energies. In all calculations the first Brillouin zone of the superlattice is sampled using a Monkhorst-Pack (MP) k-point mesh [32]. Employing the 108-atom supercell for face-centered cubic (fcc) Al we compare the results obtained with $4 \times 4 \times 4$ and $6 \times 6 \times 6$ k-point meshes to check the convergence of the total energy. Differences in the total energy of the systems are less than 5 meV per atom. All the calculations have thereafter been performed with the finer MP-mesh. A plane-wave cutoff of 300 eV is used in the calculation of the pseudo valence wave functions.

In the defect calculations, atomic positions are relaxed and the total energy is minimized until the forces acting on atoms are less than 0.04 eV/Å. The volume relaxation is not performed systematically, because we found that the use of a larger (108 and more atoms) supercells gives well-converged results without volume relaxation. However, results of test calculations employing the smallest (64 atom) supercell are given below. In the plane wave calculations we have used lattice constants optimized for each set of computational parameters, i.e., the cut-off energy and MP-mesh, used. For error cancellation, the total energy differences and relative ionic relaxations between defect and perfect bulk systems are calculated from results for supercells of the same size and obtained with the same computational parameters.

3. Results

3.1. Reliability of Modeling

To confirm the reliability of the employed numerical methods, we have calculated the formation energies (formation enthalpies at zero pressure) of mono- and di-vacancies in fcc Al. In the case of di-vacancies, we have considered nearest (1NN) and next nearest neighbor (2NN) configurations. The mono-vacancy formation energy is calculated as

$$H_V^F = E_{V(N-1)} - \frac{N-1}{N} E_{Al_{bulk}}(N) \tag{1}$$

where N is the number of atoms in the supercell, $E_{V(N-1)}$ is the total energy of a fcc Al-supercell containing a mono-vacancy, and $E_{Al_{bulk}}(N)$ is the total energy of a perfect fcc Al-supercell.

The formation energies obtained are given in Table 1. Our VASP calculations for the isolated mono-vacancy lead to values in close agreement with previous similar LDA calculations by Carling et al. [15] but also with a different approach like SIESTA [33] giving for the formation energy $H_V^F = 0.66$ eV [34,35]. All the calculated values deviated less than 0.05 eV from reliable experimental values [16]. Here, reference [16] is a summary of a few dozen experimental works, where the data are weighted according to their relevance by experts in the field. So, the given value of $H_V^F = 0.67$ eV for the vacancy formation energy in pure aluminum is an average of the most reliable values published.

Our results for the di-vacancy binding energies $H_{2V,X}^F - 2H_V^F$ (X = 1NN or 2NN) show that the interaction between nearest neighbor vacancies (X = 1NN) in Al is repulsive. This is in agreement with the results by Carling et al. [15] and also with the experimental finding that Al does not, in contrast to other metals like Cu [36,37], show a tendency for forming vacancy clusters after low-temperature irradiation and subsequent annealing [38]. However, the 2NN di-vacancy shows a tiny binding which is also in agreement with the results by Carling et al. [15].

Table 1. Comparison with experimental results: Formation energies for mono- H_V^F and di-vacancies $H_{2V,X}^F$ in the nearest neighbor (X = 1NN) and next nearest neighbor (X = 2NN) positions in fcc Al. The binding energies $H_{2V,X}^b$ of the two vacancies in the two configurations are also given. Positive and negative binding energies indicate repulsion and binding, respectively. (SIESTA results: see [34]).

Method	Volume Relax	MP-Mesh	Atoms	H_V^F (eV)	$H_{2V,1NN}^F$ (eV)	$H_{2V,2NN}^F$ (eV)	$H_{2V,1NN}^b$ (eV)	$H_{2V,2NN}^b$ (eV)
VASP-LDA	yes	$6 \times 6 \times 6$	64	0.71	—	—	—	—
VASP-LDA	no	$6 \times 6 \times 6$	64	0.713	1.506	1.409	+0.081	−0.016
VASP-LDA	no	$6 \times 6 \times 6$	108	0.714	1.489	1.421	+0.061	−0.007
VASP-GGA	no	$6 \times 6 \times 6$	108	0.66	—	—	—	—
SIESTA-DZP	no	$3 \times 3 \times 3$	108	0.64	—	—	—	—
Exp. [16]	—	—	—	0.67	—	—	—	—

To further check the reliability of our calculations in terms of supercell sizes and **k**-point meshes, we have computed the solubility enthalpy of Cu in Al. It is calculated as

$$\Delta H_{mix} = E_{CuinAl} - \left[\frac{N-1}{N} E_{Al_{bulk}} + \frac{1}{N} E_{Cu_{bulk}} \right] \tag{2}$$

where E_{CuinAl} is the total energy of an fcc Al-supercell containing one Cu-atom, while $E_{Al_{bulk}}$ and $E_{Cu_{bulk}}$ are the total energies of pure fcc Cu- and Al-supercells, respectively. Supercells of the same size are used to calculate these energies. Note also that positive and negative values represent endothermic and exothermic reactions, respectively.

The numbers given in Table 2 are in agreement with the first-principles results by Wolverton et al. [17]. The small deviations indicate that, at least for single isolated Cu atoms, the energetics is well converged already for a supercell size of 108 atoms.

Table 2. Solubility enthalpy ΔH_{mix} of Cu in Al calculated by using Equation (2) and different supercell sizes.

Supercell Size (atom)	ΔH_{mix} (meV)
108	−50.5
128	−54.2
144	−53.0

3.2. Impurity-Cluster Binding Energies

The reliability of our calculations for Cu clusters is based on the tests described above. While for VASP the transferability of the pseudo potentials is well established, this is not the case for other methods. Especially, for SIESTA [33] the employed pseudo potentials have to be tested in well-known Al-Cu binding configurations of Al_2Cu as in ref. [34].

To begin with, we give in Table 3 the binding energies for a pair of Cu atoms with respect to two separate Cu atoms. These energies are obtained by optimizing the lattice constant for Al for each supercell size and **k**-point mesh. All the supercell sizes lead to a binding energy of around 50 meV. For the 128 atom supercell the binding energy is the smallest one reflecting the small spacing between the adjacent Cu habit planes of the periodic images and the ensuing artificial interaction.

Table 3. Binding energy of two Cu solute atoms in Al on nearest neigbor positions in fcc Al. The c-direction is perpendicular to the habit plane of the Cu atoms. Negative signs indicate binding.

Scheme	Number Atoms	Size Unit Cells	k-points	E_{bind} (meV)
LDA	108	$3 \times 3 \times 3$	$4 \times 4 \times 4$	−50.3
LDA	128	$4 \times 4 \times 2$	$4 \times 4 \times 8$	−46.3
LDA	144	$3 \times 3 \times 4$	$4 \times 4 \times 4$	−56.2
LDA	192	$4 \times 4 \times 3$	$3 \times 3 \times 6$	−54.7
GGA	108	$3 \times 3 \times 3$	$6 \times 6 \times 6$	−51.5

We will give the total binding energies of two- (2D) and three-dimensional (3D) copper clusters as the energy gain relative to separated Cu atoms in aluminum. From this we calculate the binding energy also per Cu-atom, i.e., the average over the cluster. We calculate also the binding energy of the last Cu-atom attached to a cluster, which indicates, if it is energetically favorable for an already existing cluster to grow further by attaching another Cu atom. This energy has to be compared to the thermal energy at room temperature of $3/2\, kT = 40$ meV. The construction scheme of the 2D Cu-platelets on the {100}-plane of the fcc Al-lattice is shown in Figure 1. It is based on well-established experimental facts on Cu-platelet formation [3,5,6,39].

Firstly, we have performed calculations on small agglomerates up to 4 Cu atoms. The results are presented in Table 4. We observe that two Cu-atoms on the 1NN positions are bound together with a binding energy of about 50 meV, while there is a weak binding of 2NN Cu-atoms of about 10 meV as well. However, the most important result is that 2D agglomerates of 4 Cu on the {100}-plane of the fcc Al are preferred instead of the 3D tetrahedron structure (h) with an energy difference of 259 meV for the triangle configuration (g) and with 398 meV for the rectangle configuration (f) in Table 4 (cf. also Figure 1).

Table 4. Binding energies for agglomerates of Cu atoms in 1D, 2D, and 3D configurations. The calculation employed the 108 atom supercell. Given is the total binding energy, the binding energy per Cu-atom, and the binding energy of the 'last' Cu atom specified in Figure 1. We give here the energy with an accuracy of 0.1 meV, which is only of internal numerical relevance. The numbering is according to Figure 1 left.

Agglomerate	atom no.	Spatial	E_{bind} (meV)	E_{bind} (meV)	E_{bind} (meV)
structure		extension	cluster	per Cu	last Cu
(a) 2 Cu 1NN on (100)-plane	1, 2	2D	−50.3	−25.1	−50.2
(b) 2 Cu 2NN on (100)-plane	1, 4	2D	−9.6	−4.8	−9.6
(c) 3 Cu in-line on (100)-plane	6, 7, 8	1D	−95.2	−31.7	−45.0
(d) 3 Cu triangle on (100)-plane	1, 2, 3	2D	−134.7	−44.9	−84.4
(e) 3 Cu triangle on (111)-plane	–	2D	−97.4	−32.5	−47.1
(f) 4 Cu rectangle on (100)-plane	1, 2, 3, 4	2D	−344.7	−86.2	−210.0
(g) 4 Cu triangle on (100)-plane	1, 2, 3, 5	2D	−206.1	−51.5	−71.4
(h) 4 Cu tetrahedron in space	–	3D	+53.0	+13.2	+150.4

Experimentally, there is, since the early investigations by Guinier and Preston [5,6] a long-standing agreement on the fact that Cu prefers to precipitate as 2D platelets on the {100}-planes of the fcc Al [40,41] . Recently, this has been even confirmed by HR-TEM [39] showing mono-atomic platelets on the three equivalent {100}-planes in the fcc lattice. Our results presented in Table 4 confirm this perception. Of the triangular structures (d) and (e) the platelet structure (d) in the (100)-plane is favored by nearly 40 meV compared to the platelet structure (e) on a (111)-plane. However, the close-packed structure (h) of Cu-atoms in the form of a 3D tetrahedron even shows repulsion. Please note that larger Cu-platelets ((f) in Table 4) show the largest binding per Cu-atoms, i.e., they are the most stable ones.

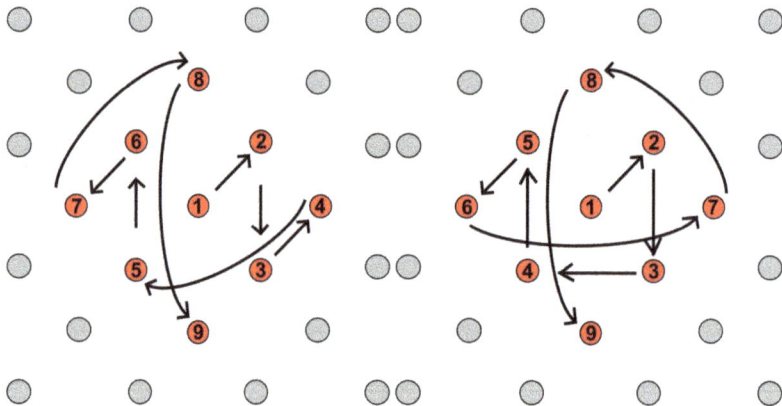

Figure 1. Configuration of Cu-atoms in 2D platelets on the {100}-plane in fcc Al. The Cu and Al atoms are shown by red and grey spheres, respectively. The numbering indicates how the Cu-platelets were assumed to grow. The left and right patterns show the sequences 1 and 2 used in the calculations, respectively.

3.2.1. 2D-Cu-Clusters in 108-atom Supercells

Having confirmed the preference for copper to precipitate as platelets on the {100}-planes in fcc aluminnum, we have constructed larger platelets starting from the two different 4-Cu-atom configurations (f) and (g) in Table 4 and using the sequences 1 and 2 in Figure 1, respectively. The numbering gives the order, in which the Cu-atoms have been attached to the Cu-platelet. All calculations have been performed in the LDA scheme using a MP-mesh given in Table 5.

The binding energy of the last Cu atom to a cluster of $N-1$ atoms is shown in the left graph of Figure 2, where $n = 2$ corresponds to the Cu-atom pair on 1NN positions. The way of constructing the Cu-platelets is shown in Figure 1. From Figure 2 it is evident that the triangular structure of four Cu atoms (sequence 2) is much less favorable than the square one (sequence 1). This means that once the quite stable triangle has formed, the forth copper atom attaching is likely to complete the small triangle rather to a square (sequence 1) than forming a larger triangle (sequence 2). This is also reflected in the higher binding energy per Cu atom shown in the right graph of Figure 2. Attaching the 7th, 8th, or 9th atom does not make a difference between the sequences due to their symmetry.

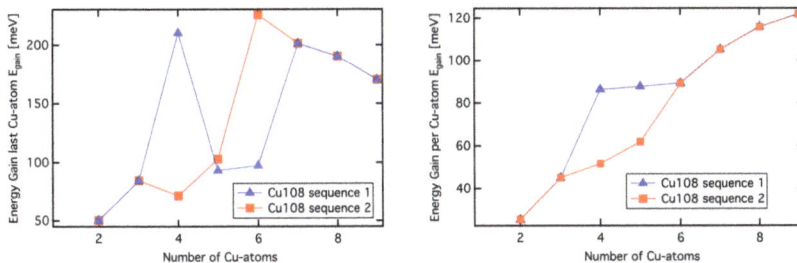

Figure 2. Energy gain during Cu-cluster growth for a supercell of 108 atoms in sequences 1 and 2. (**left**) The energy gain due to the last attached Cu atom, (**right**) the total energy gain per Cu atom in the growing cluster. Sequences 1 and 2 are explained in Figure 1.

Table 5. Used supercells with corresponding MP-meshes.

Type	Size (atoms)	Size (Unit Cell)	MP-mesh
standard	108	$3 \times 3 \times 3$	$4 \times 4 \times 4$
flattened	128	$4 \times 4 \times 2$	$4 \times 4 \times 6$
elevated	144	$3 \times 3 \times 4$	$3 \times 3 \times 2$
widened	192	$4 \times 4 \times 3$	$2 \times 2 \times 3$

The right part of Figure 2 shows the energy gain per copper atom for platelets consisting of 4- and 5-atom in sequence 1 in comparison to sequence 2. This significant increase is caused by relaxations perpendicular to the habit plane of the Cu atoms.

3.2.2. 2D-Cu-Clusters in 128- and 192-atom Supercells

Lattice relaxations within the habit plane of the Cu-platelet cause an interaction between neighboring supercells. Hence, a 108-atom supercell ($3 \times 3 \times 3$ unit cells) is already quite small to accommodate a Cu-platelet consisting of more than 5 atoms. Thus, we have repeated some of the calculations in enlarged supercells to keep the growing platelets farther apart from each other between the periodic images of the supercells.

On the one hand, we employed a supercell of 128 atoms ($4 \times 4 \times 2$ unit cells) increasing the lateral distance of the periodic images, while reducing the distance perpendicular to the platelets in the $< 001 >$-direction. On the other hand, a supercell of 192 atoms ($4 \times 4 \times 3$ unit cells) increases the lateral distance while keeping the distance in between Cu-platelets in the $< 001 >$-direction the same as for the 108-atoms cell (cf. Table 5). The results for these supercells are shown in Figures 3 and 4.

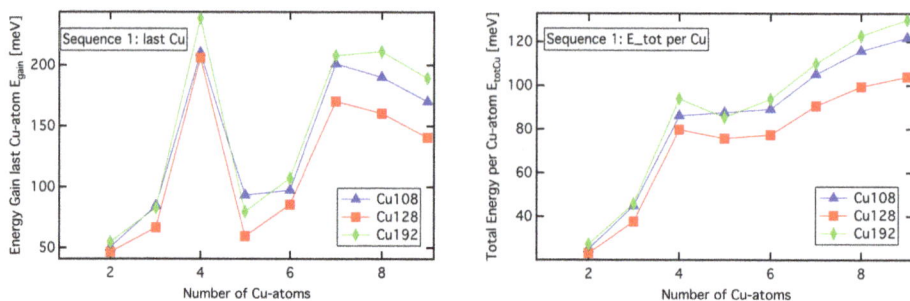

Figure 3. Energy gain during Cu-cluster growth. Results corresponding to supercells of 108, 128, and 192 atoms are compared. (**left**) Energy gain due to the last Cu atom attached, and (**right**) total energy gain per Cu-atom of a growing cluster. Sequence 1 is explained in Figure 1.

For the flattened supercell of 128 atoms (see Table 5) the separation between the copper platelets in the habit plane increases compared to the supercell of 108 atoms, while the distance between the platelets and their periodic images becomes obviously too small. Figures 3 and 4 show that the energy gain by relaxation is significantly smaller for the supercell of 128 atoms, which artificially suppresses the energy-lowering relaxation perpendicular to the platelets (see Figure 5).

The use of the supercell of 144 atoms, on the other hand, just increases the separation perpendicular to the copper platelets (see Table 5). Thus, the size of the platelets has to be limited to five Cu atoms and therefore this configuration is not considered further. Finally, the use of the supercell of 192-atoms (4×4 lattice constants wide, but three lattice constants in height as well) reduces the interactions between copper platelets in the habit plane, while the separation of the platelets in the $< 001 >$-direction is the same as for the supercell of 108 atoms. Thus, this supercell gives the largest energy gain as seen from Figures 3 and 4.

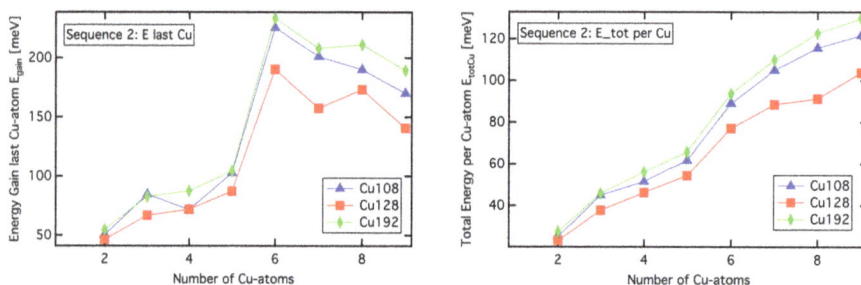

Figure 4. Energy gain during Cu-cluster growth. The results corresponding to the supercells of 108, 128, and 192 atoms are compared. (**left**) Energy gain of the last Cu atom attached, (**right**) total energy gain per Cu-atom of a growing cluster. Sequence 2 is explained in Figure 1.

Since the energy differences are nevertheless small, i.e., less than 10% between the supercells of 108 and 192 atoms, we have chosen the computationally more feasible supercell of 108 atoms for the following discussion.

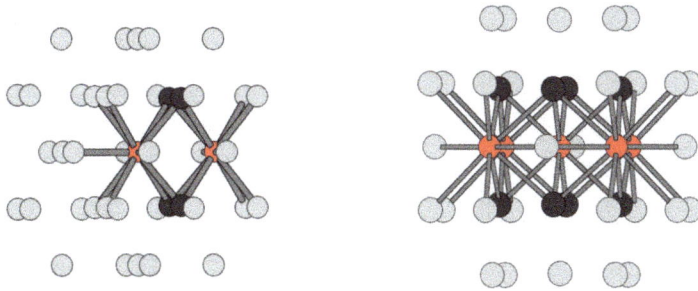

Figure 5. Relaxation patterns of growing clusters. (**left**) Two copper atoms (red spheres) on 1NN position, (**right**) five Cu-atoms arranged as a platelet on the {100}-plane. The Al atoms having two or more bonds to Cu-atoms are plotted in dark-grey color. The relaxation of the Al-layer above and below towards the Cu-atoms on the {100}-plane clearly increases with the number of agglomerated Cu-aotms.

3.2.3. Lattice Relaxations

As shown in Figure 5, the relaxation of the Al-layers above and below a copper platelet, residing on the {100}-plane, clearly increases with the number of agglomerated copper atoms. For a platelet consisting of five Cu atoms the aluminum layers relax towards the copper platelet by about 20 pm or 10% compared to the ideal separation of lattice planes (202 pm). This is a similar behavior as observed numerically for 5-atomic copper platelets [34] or experimentally [5,40,41] for GP-I zones, which are described as a single extended layer of Cu atoms on a {100}-plane (see [3]).

3.2.4. Relaxed Versus Static Configurations

To clarify the role of the strong relaxation around the Cu atoms in the aluminum lattice on the observed energy gain, we compare static (atoms fixed at their ideal aluminium lattice positions) and relaxed atomic configurations. Figure 6 shows the results for a supercell of 108 atoms and for 2D and 3D configurations. Even though the unrelaxed case is unphysical, it gives already the major contribution to the energy gain in forming Cu-platelets. We can also conclude that the energy gain due to relaxation of the surrounding Al atoms is more important than the exact shape of the precipitate.

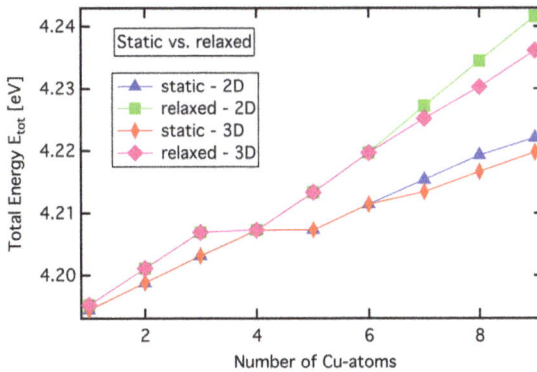

Figure 6. Total energies calculated by using static (atoms fixed) and relaxed configurations of a supercell of 108 atoms.

3.2.5. Comparison with Zinc Clusters

In contrast to copper, which has an fcc structure, zinc crystalizes in a hexagonal closed-packed (hcp) lattice. Our calculations show that there is a binding energy between the 1NN Zn-atoms. However, it is clearly smaller than in the case of copper—i.e., only around 20 meV. The binding energies of Zn atom clusters in different 2D platelet and 3D configurations are given in Figure 7. In the 3D configuration with five Zn atoms the binding energy is around 24 meV/Zn atom whereas in the 2D configuration it is only around 14 meV/Zn atom. This preference to form 3D precipitates is in contrast to the behavior of Cu precipitates and it reflects the different lattice structures of copper and zinc resulting in the tendency of Zn to form spherical precipitates [42–45].

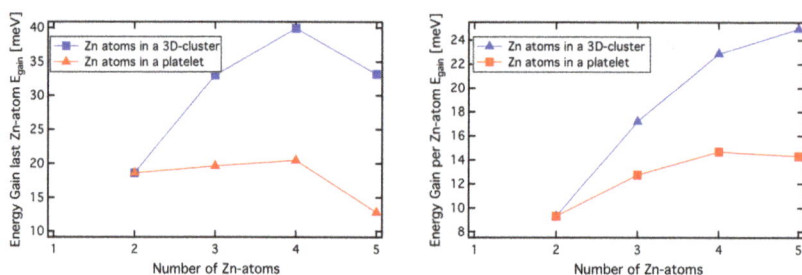

Figure 7. Binding energies of Zn atoms in 2D and 3D Zn clusters. (**left**) The binding energy for the last attached Zn atom. (**right**) The binding energy per Zn atom in the cluster.

3.2.6. Pre-Guinier-Preston Zones in Al-Cu and Al-Zn

The small energetically-favored 2D Cu precipitates on the Al {100}-planes, shown schematically in Figure 1 or with relaxed atomic positions in Figure 5, can be considered to be the starting point of growing Guinier-Preston zones. Guinier-Preston zones become visible in XRD when they have reached a size of about 1–1.4 nm [41], i.e., consisting of more than 24–48 atoms. A size of 1.4 nm may be still too small for visibility in the TEM or HR-TEM. However, due to the computational limitations we could consider here only platelets of the maximal number of nine atoms.

Due to the experimental fact that detects in as-quenched pure Al-Cu alloys show positron lifetimes, which should be related to an open volume of about a mono-vacancy in Al [4,46] one can assume that—at least some—GP zones must contain structural vacancies. Some structures of Cu-platelets containing vacancies have been calculated by SIESTA in ref. [34].

4. Discussion and Conclusions

Our computational results employing DFT-LDA are consistent and in sufficient agreement with previously calculated and experimental values. Especially, our results on vacancies in aluminum agree well with other ab initio calculations [15] and experiments [16] giving credence to our approach. Moreover, the inward relaxation of nearest-neighbor Al-atoms surrounding a single Cu atom is in accordance with experimental results from X-ray absorption [47]. Furthermore, we find a binding between Cu atoms situated on the {100}-habit planes in fcc aluminum, while, e.g., a 3D tetrahedron even shows a repulsive interaction. Thus, 2D structures are favored compared to 3D ones, which is in agreement with experimental observations finding mono-atomic copper platelet on the {100}-planes in the fcc Al-lattice when segregation of the super saturated solution is starting [41].

The calculated total energies for a 2D copper cluster vary slightly with used supercell sizes and shapes. However, the dependence of the total energy on the number of Cu atoms shows a similar trend in all cases. Nevertheless, it turns out that there have to be at least three layers of Al atoms in the supercell perpendicular to the Cu-habit plane separating periodic images of the copper platelets. Obviously, the relaxation of aluminum planes parallel to the copper platelet, which is important for

Appl. Sci. **2018**, *8*, 1003

energy lowering, cannot be reliably described with wide but flat supercells. This type of cells is not large enough in the z-direction perpendicular to the copper platelet.

In contrast to magic number arrangement of defects in semiconductors [48], for metallic materials it seems to be difficult to state that distinguished structures are existing at all. While in general the binding energies for different geometric configurations differ very little (about 50–100 meV), we also find some special platelet structures of Cu atoms with quite high binding energies between the copper species. However, the generally small differences in binding energies may be responsible for not finding any special configurations in experiments. Remarkable is only that some closed configurations of Cu atoms on the {100}-planes show higher binding energies per Cu atom (four Cu arranged in a square). Obviously, their geometries allow electronic structures, which are energetically more favorable. However, it seems to be experimentally demanding to confirm this finding.

Concerning 3D instead of 2D plate-like structures, such as the 3D-tetrahedron, we find that these close-packed structures are the least favorable ones at all, since the Cu atoms inside their habit plane have a tendency to move away from each other. Hence, close-packed 3D-structures are unlikely to be energetically favorable in Al-Cu alloys.

According to our calculations the formation of platelets during the aging of Al-Cu alloys is in contrast to Al-Zn alloys, where obviously 3D-Clusters are preferred. Again, this is in accordance with experimental results for pure Al-Zn alloys, where clusters grow in early stages of decomposition in a spherical manner [41,43,45].

Author Contributions: Conceptualization, P.F. and M.J.P. and T.E.M.S.; Methodology, P.F. and M.J.P.; Software, P.F.; Validation, P.F. and M.J.P. and T.E.M.S.; Formal Analysis, P.F. and T.E.M.S.; Investigation, P.F.; Resources, P.F.; Data Curation, P.F.; Writing Original Draft Preparation, P.F. and T.E.M.S.; Writing Review & Editing, M.J.P. and I.W.; Visualization, T.E.M.S.; Supervision, M.J.P.; Project Administration, M.J.P. and T.E.M.S.; Funding Acquisition, P.F. and T.E.M.S.

Funding: This research was funded in part by the German Research Foundation (Deutsche Forschungsgemeinschaft—DFG) grant number STA 527/3-2 and STA 527/5-1. This publication was funded by the German Research Foundation (DFG) and the University of Wuerzburg in the funding programme Open Access Publishing.

Acknowledgments: We acknowledge large computational resources provided by Politecnico di Milano on the clusters Avogadro.

Conflicts of Interest: The authors declare no conflict of interest. The founding sponsors had no role in the design of the study; in the collection, analyses, or interpretation of data; in the writing of the manuscript, and in the decision to publish the results.

Abbreviations

The following abbreviations are used in this manuscript:

PAS	Positron Annihilation Spectroscopy
XAFS	X-ray Absorption Fine Structure
DFT	Density Functional Theory
LDA	Local Density Approximation
GGA	General Gradient Approximation
VASP	Vienna Ab initio Simulation Package
PAW	Projector Augmented Wave
MP	Monkhorst-Pack
SIESTA	Spanish Initiative for Electronic Simulations with Thousands of Atoms
DZP	Double Zeta Polarized
1NN	Nearest Neighbors
2NN	Next Nearest Neighbors
XRD	X-Ray Diffraction
TEM	Transmission Electron Microscopy
HR-TEM	High resolution Transmission Electron Microscopy
fcc	face-centered cubic
hcp	hexagonal closed-packed

References

1. Polmear, I. *Light Alloys—From Traditional Alloys to Nanocrystals*, 4th ed.; Elsevier: Amsterdam, The Netherlands, 2006.
2. Ostermann, F. *Anwendungstechnologie Aluminium*, 3rd ed.; Springer: Berlin/Heidelberg, Germany, 2014.
3. Haasen, P. *Physical Metallurgy*, 3rd ed.; Cambridge University Press: Cambridge, UK, 1996.
4. Dupasquier, A.; Folegati, P.; de Diego, N.; Somoza, A. Current positron studies of structural modifications in age-hardenable metallic systems. *J. Phys. Condens. Matter* **1998**, *10*, 10409–10422. [CrossRef]
5. Guinier, A. La diffraction de rayon X aux tres petis angles: Application a l'etude de phenomenes ultramicroscopiques. *Ann. Phys.* **1938**, *12*, 161–237.
6. Preston, G. The Diffraction of X-Rays by an Age-Hardening Aluminum and Copper Alloys. *Proc. R. Soc. Lond. Ser. A Math. Phys. Sci.* **1938**, *167*, 526–538. [CrossRef]
7. Preston, G. The Diffraction of X-Rays by an Age-Hardening Alloy of Aluminum and Copper: The Structure of an Intermediate Phase. *Lond. Edinb. Dublin Philos. Mag. J. Sci.* **1938**, *26*, 855–871. [CrossRef]
8. De Geuser, F.; Bley, F.; Denquin, A.; Deschamps, A. Mapping the microstructure of a friction-stir welded (FSW) Al-Li-Cu alloy. *J. Phys.: Conf. Ser.* **2010**, *247*, 012034. [CrossRef]
9. De Geuser, F.; Bley, F.; Deschamps, A. A new method for evaluating the size of plate-like precipitates by small-angle scattering. *J. Appl. Cryst.* **2012**, *45*, 1208–1218. [CrossRef]
10. Decreus, B.; Deschamps, A.; Donnadieu, P.; Ehrström, J.C. On the Role of Microstructure in Governing Fracture Behavior of an Aluminum-Copper-Lithium Alloy. *Mater. Sci. Eng. A* **2013**, *586*, 418–427. [CrossRef]
11. Le Jolu, T.; Morgeneyer, T.; Denquin, A.; Sennour, M.; Laurent, A.; Besson, J.; Ourgues-Lorenzon, A.F. Microstructural Characterization of Internal Welding Defects and their Effect on the Tensile Behaviour of FSW Joints of AA2198 Al-Cu-Li Alloy. *Metall. Mater. Trans. A* **2014**, *45A*, 5531–5544. [CrossRef]
12. Bourgeois, L.; Dwyer, C.; Weyland, M.; Nie, J.F.; Muddle, B. Structure and Energetics of the Coherent Interface between the Θ' Precipitate Phase and Aluminium in Al-Cu. *Acta Mater.* **2011**, *59*, 7043–7050. [CrossRef]
13. Zhang, Y.; Zhang, Z.; Medhekar, N.; Bourgeois, L. Vacancy-tuned Precipitation Pathways in Al-1.7 Cu-0.025In-0.025Sb (at.%) Alloy. *Acta Mater.* **2017**, *141*, 341–351. [CrossRef]
14. Carling, K.; Wahnström, G.; Mattsson, T.; Sandberg, N.; Grimvall, G. Vacancy Concentration in Al from Combined first-principles and Model Potential Calculations. *Phys. Rev.* **2003**, *B67*, 054101. [CrossRef]
15. Carling, K.; Wahnström, G.; Mattsson, T.; Mattsson, A.; Sandberg, N.; Grimvall, G. Vacancies in Metals: From first-principles Calculations to Experimental Data. *Phys. Rev. Lett.* **2000**, *85*, 3862–3865. [CrossRef] [PubMed]
16. Ullmaier, H. (Ed.) *Landolt-Börnstein—Numerical Data and Functional Relationships in Science and Technology—New Series—Group III: Crystal and Solid State Physics Volume 25: Atomic Defects in Metals*; Springer: Berlin, Germany, 1991.
17. Wolverton, C.; Ozoliņš, V. First-principles Aluminum Database: Energetics of Binary Al Alloys and Compounds. *Phys. Rev.* **2006**, *B73*, 144104. [CrossRef]
18. Wolverton, C. Solute-Vacancy Binding in Aluminum. *Acta Mater.* **2007**, *55*, 5867–5872. [CrossRef]
19. Shin, D.; Wolverton, C. First-principles Study of Solute-Vacancy Binding in Magnesium. *Acta Mater.* **2010**, *58*, 531–540. [CrossRef]
20. Haaks, M.; Staab, T. High Momentum Analysis in Doppler Spectroscopy. *Appl. Surf. Sci.* **2008**, *255*, 84–88. [CrossRef]
21. Klobes, B.; Staab, T.; Haaks, M.; Maier, K.; Wieler, I. The Role of Quenched-in Vacancies for the Decomposition of Aluminium Alloys. *Phys. Status Solidi (RRL) Rapid Res. Lett.* **2008**, *2*, 224–226. [CrossRef]
22. Klobes, B.; Korff, B.; Balarisi, O.; Eich, P.; Haaks, M.; Maier, K.; Sottong, R.; Hühne, S.M.; Mader, W.; Staab, T. Probing the Defect State of Individual Precipitates Grown in an Al-Mg-Si Alloy. *Phys. Rev.* **2010**, *B82*, 054113. [CrossRef]
23. Klobes, B.; Maier, K.; Staab, T. Natural Ageing of Al-Cu-Mg Revisited from a Local Perspective. *Mater. Sci. Eng. A* **2011**, *528*, 3253–3260. [CrossRef]

24. Klobes, B.; Staab, T.; Dudzik, E. Early Stages of Decomposition in Al Alloys Investigated by X-Ray Absorption. *Phys. Status Solidi (RRL) Rapid Res. Lett.* **2008**, *2*, 182–184. [CrossRef]
25. Klobes, B.; Korff, B.; Balarisi, O.; Eich, P.; Haaks, M.; Kohlbach, I.; Maier, K.; Sotong, R.; Staab, T. Defect Investigations of Micron Sized Precipitations in Al Alloys. *J. Phys. Conf. Ser.* **2011**, *262*, 012030. [CrossRef]
26. Wang, S.; Starink, M. Precipitations and Intermetallic Phases in Precipitation Hardening Al-Cu-Mg-(Li) Based Alloys. *Int. Mater. Rev.* **2005**, *50*, 193–215. [CrossRef]
27. Staab, T.; Zamponi, C.; Haaks, M.; Modrow, H.; Maier, K. Atomic Structure of pre-Guinier-Preston Zones in Al-Alloys. *Phys. Status Solidi (RRL) Rapid Res. Lett.* **2007**, *1*, 172–174. [CrossRef]
28. Makkonen, I.; Hakala, M.; Puska, M. Calculation of Valence Electron Momentum Densities using the Projector Augmented-Wave Method. *J. Phys. Chem. Solids* **2005**, *66*, 1128–1135. [CrossRef]
29. Rehr, J.; Albers, R. Theoretical Approaches to X-Ray Absorption Fine Structure. *Rev. Mod. Phys.* **2000**, *72*, 621–654. [CrossRef]
30. Kresse, G.; Furthmüller, J. Efficiency of *ab initio* Total Energy Calculations for Metals and Semiconductors using a Plane-Wave Basis Set. *Comput. Mater. Sci.* **1993**, *6*, 15–50. [CrossRef]
31. Kresse, G.; Furthmüller, J. Efficient Iterative Schemes for *ab initio* Total-Energy Calculations using a Plane-Wave Basis Set. *Phys. Rev.* **1996**, *B54*, 11169–11186. [CrossRef]
32. Monkhorst, H.; Pack, J. Special Points for Brillouin-Zone Integrations. *Phys. Rev.* **1976**, *B13*, 5188–5192. [CrossRef]
33. Soler, J.; Artacho, E.; Gale, J.; García, A.; Junquera, J.; Ordejón, P.; Sánchez-Portal, D. The SIESTA Method for ab initio Order-N Materials Simulation. *J. Phys. Condens. Matter* **2002**, *14*, 2745–2779. [CrossRef]
34. Kohlbach, I.; Korff, B.; Staab, T. (Meta-) Stable Phases and Pre-Guinier-Preston Zones in AlCu-Alloys Constructed from ab initio Relaxed Atomic Positions—Comparison to Experimental Methods. *Phys. Status Solidi B* **2010**, *247*, 2168–2178. [CrossRef]
35. Wolfertz, I. Ab initio Untersuchungen an frühen Ausscheidungsphasen der Aluminium(-Magnesium-) Kupfer-Legierungen. Ph.D. Thesis, HISKP, University Bonn, Bonn, Germany, 2014.
36. Mantl, S.; Triftshäuser, W. Defect Annealing Studies on Metals by Positron Annihilation and Electrical Resistivity Measurements. *Phys. Rev.* **1978**, *B17*, 1645–1652. [CrossRef]
37. Staab, T.; Krause-Rehberg, R.; Vetter, B.; Kieback, B. The influence of Microstructure on the Sintering Process in Crystalline Metal Powders investigated by Positron Lifetime Spectroscopy: Part I: Electrolytic and Spherical Copper Powder. *J. Phys. Condens. Matter* **1999**, *11*, 1757–1786. [CrossRef]
38. Rajainmäki, H.; Linderoth, S. Stage II Recovery in Proton-Irradiated Aluminum Studied by Positrons. *J. Phys. Condens. Matter* **1990**, *2*, 6623–6630. [CrossRef]
39. Konno, T.; Kawasaki, M.; Hiragi, K. Characterization of Guinier-Preston Zones by High-Angle Annular Detector Dark-Field Scanning Transmission Electron Microscopy. *JEOL News* **2001**, *36E*, 14–17.
40. Hardy, H. Report on Precipitation. *Prog. Met. Phys.* **1954**, *5*, 143–278. [CrossRef]
41. Baur, R.; Gerold, V. Entmischungsvorgänge im System Aluminium-Kupfer. *Zeitschrift für Metallkunde* **1966**, *57*, 181–186.
42. Gerold, V.; Schweizer, W. Die Kinetik von Entmischungsvorgängen in übersättigten Aluminium-Zink-Mischkristallen. *Zeitschrift für Metallkunde* **1961**, *52*, 76–86.
43. Gerold, V. Die Zonenbildung in Aluminium-Zink-Legierungen. *Phys. Status Solidi A* **1961**, *1*, 37–49. [CrossRef]
44. Dlubek, G.; Kabisch, O.; Brümmer, O.; Löffler, H. Precipitation and Dissolution Processes in Age-Hardenable Al Alloys—A Comparison of Positron Annihilation and X-Ray Small Angle Scattering Investigations—I. Al-Zn(x) (x = 3, 4.5, 6, 10, 18 at.%). *Phys. Status Solidi A* **1979**, *55*, 509–518. [CrossRef]
45. Krause, R.; Dlubek, G.; Wendrock, G. Structural Changes during Post-Ageing of an Al-Zn (15 at.%) Alloy at 100°C Studied by Positron Annihilation, Small Angle X-Ray Scattering and Microhardness Measurements. *Cryst. Res. Technol.* **1985**, *20*, 1495–1501. [CrossRef]
46. Gläser, U.; Dlubek, G.; Krause, R. Vacancies and Precipitates in Al—1.9 at % Cu Studied by Positrons. *Phys. Status Solidi A* **1991**, *163*, 337–343. [CrossRef]

47. Fontaine, A.; Lagarde, P.; Naudon, A.; Raoux, D.; Spanjaard, D. EXAFS Studies on Al-Cu Alloys. *Philos. Mag. B* **1979**, *40*, 17–30. [CrossRef]
48. Staab, T.; Haugk, M.; Frauenheim, T.; Leipner, H. Magic Number Vacancy Clusters in GaAs—Structure and Positron Lifetime Studies. *Phys. Rev. Lett.* **1999**, *83*, 5519–5522. [CrossRef]

*applied
sciences*

MDPI

Article

Effects of Heat Treatment on the Tribological Properties of Sicp/Al-5Si-1Cu-0.5Mg Composite Processed by Electromagnetic Stirring Method

Ning Li [1,2], Hong Yan [1,*] and Zhi-Wei Wang [1]

[1] School of Mechanical Electrical Engineering, Nanchang University, Nanchang 330031, China;
    lining@nchu.edu.cn (N.L.); wangzhiw@landwind.com (Z.-W.W)
[2] School of Aviation Manufacturing Engineering, Nanchang Hangkong University, Nanchang 330069, China
* Correspondence: hyan@ncu.edu.cn; Tel.: +86-791-8396-9633; Fax: +86-791-8396-9622

Received: 29 January 2018; Accepted: 1 March 2018; Published: 4 March 2018

Abstract: This paper investigated the influence of heat treatment (T6) on the dry sliding wear behavior of SiC_p/Al-5Si-1Cu-0.5Mg composite that was fabricated by electromagnetic stirring method. The wear rates and friction coefficients were measured using a pin-on-disc tribometer under loads of 15–90 N at dry sliding speeds of 100 r/min, 200 r/min, and 300 r/min, over a sliding time of 15 min. The worn surfaces and debris were examined using a scanning electron microscope and was analyzed with an energy dispersive spectrometer. The experimental results revealed that SiC_p/Al-5Si-1Cu-0.5Mg alloy treated with T6 exhibited lower wear rate and friction coefficient than the other investigated alloys. As the applied load increased, the wear rate and friction coefficient increased. While, the wear rate and friction coefficient decreased with the sliding speed increasing. The morphology of the eutectic silicon was spheroidal after the T6 heat treatment. SiC_p particles and Al_2Cu phase can be considered as the main raisons for improving the wear behavior. Abrasion and oxidation were the wear mechanisms at low load levels. However, the wear mechanisms at high load levels were plastic deformation and delamination.

Keywords: heat treatment; SiC_p/Al-5Si-1Cu-0.5Mg composite; wear resistance; wear mechanisms

---

1. Introduction

In recent years, Al-Si alloys have been widely used in the manufacture of vehicles, due to their excellent physical properties, including low desity, high strength-to-weight ratio, good fluidity, low coefficient of thermal expansion and good machinability [1]. However, under the conditions of the traditional casting process, Al-Si alloys usually contain massive coarse, long-needle or lamellar shape eutectic Si phases. Those defects often lead to poor mechanical properties and wear resistance.

The addition of rare earth (RE) [2], Sr [3], Al_2O_3 [4], or SiC [5,6], and heat treatment [7], are effective ways to change the morphology and distribution of the eutectic silicon phase in the Al matrix to improve the mechanical properties of the alloy. Xiaofan Du [5] studied the effect of situ synthesizing SiC particles on the Al-Si alloy. An in situ 2% SiC_p reinforced Al-Si alloy was designed. The hardness and wear resistance were improved. Besides, it was regarded that the in situ SiC particulates act on the process of heat treatment, which affected the mechanical properties as a result. But, the samples were prepared by traditional method and the wear mechanisms were not analysed. Rajeev et al. [6] studied the wear behavior of the Al-Si-SiCp composites by using a reciprocating friction wear test. This study showed that nomal load, reciprocating velocity, sliding distance, and silicon content all had a significant effect on the wear property of the samples. Among these factors, it was found that the increased Si content altered the composites' wear resistance. On the other hand, Gupta et al. [8] studied the microstructures of the Al-Si alloy and found that they were altered by T6 heat treatment. T6 heat treatments increased the hardness, strength, elongation, and wear resistance of the alloy. Singh J et al. [9] reported the influence of wear test parameters on the wear performance of Al-composites of mechanical parameteres, such as: applied load, sliding velocity, sliding distance, temperature, and counterface hardness. The results revealed that these parameters can influence the wear and tribology behaviors of Al-composites in dry sliding wear tests.

Recently, the process of non-contacting method using an electromagnetic field is being conducted. When compared to traditional stirring casting, this method has the advantages of nonpollution, easy process control, and continuous production [10]. It has been a main method for producing the semi-solid slurry or billets. Therefore, electromagnetic stirring (EMS) is one of the most promising methods for processing cast products. EMS with optimal parameters has a remarkable ability to improve the macrosopic quality of billet [11]. Stirring power is a key process factor and has a direct and positive influence on microstructural evolution [12–14]. The suitable slight electromagnetic stirring that is applied to the melt during the low superheat pouring can increase the number of crystal nuclei in the melt to reduce the grain size, and can make the primary phase morphology to become more round. Dwivedi [15] believed that this process resulted in the distribution of SiC particles evenly with a low porosity. Dwivedi found that SiCp were evenly distributed in the matrix due to the melt flow that is induced by the rotating magnetic field of electromagnetic stirring. This indicates the effectiveness of the technique electromagnetic stir casting process that is utilized for the production. In a word, the EMS method can improve the mechanical properties and microstructural composition of the alloy, and may improve the wear resistance of the alloy.

Few literatures can be found to reveal the effect of T6 heat treatment on the wear properties of Al-Si alloys that are reinforced with SiCp processed by electromagnetic stirring method. T6 heat treatment refers to a solution treatment at 520–550 °C for several hours, followed by quenching and artificial ageing [16]. In the present work, Al-5Si-1Cu-0.5Mg alloy, which is hypoeutectic Al-Si alloy, was adopted as the matrix. An electromagnetic stirring technique was used to prepare the SiCp/Al-5Si-1Cu-0.5Mg composite. After T6 heat treatment, the microstucture, tensile strength, hardness and elongation of the composite were tested. A pin-on-disc test was also carried on to investigate the heat treatment on the wear properties of SiCp/Al-5Si-1Cu-0.5Mg composite processed by electromagnetic stirring method. The wear mechanisms of the tests were analyzed by scanning electron microscope (SEM) and energy dispersive spectroscopy (EDS).

2. Experimental

2.1. Materials Preparation

The addition of SiCp reinforcement to a metal matrix provides considerably better wear and thermal properties [17]. Therefore, SiCp particles with an average size of about 2 um were chosen as the reinforcement particles. The SEM image of the received SiCp particles is shown in Figure 1. The SiCp was pretreated by high temperature oxidation (oxidation for 4 h under 1000 °C). Because the SiCp without pretreatment is prone to forming a harmful, brittle Al_4C_3 phase when it directly comes into a molten Al matrix. The high temperature oxidation also improved the wettability of SiCp and the molten Al.

An electromagnetic stirrer (EMS-SM05) was used for the preparation of the materials. Figure 2 shows a schematic diagram of the electromagnetic stirring set-up. The electromagnetic frequency was 35 Hz and the current was 57.5 A. The stirring nominal power was 20,125 W.

Figure 1. SEM (scanning electron microscope) image of the received SiC particles.

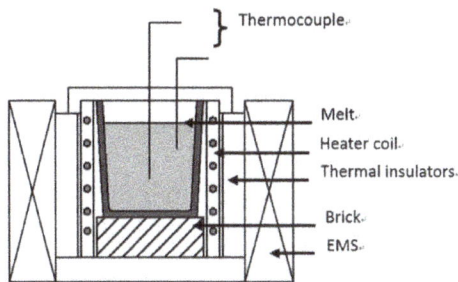

Figure 2. The diagram of electromagnetic stirrer. EMS: electromagnetic stirring.

Ingots of Al-5Si-1Cu-0.5Mg alloy were used as the matrix alloy in this research, and the compositions of this alloy were listed in Table 1. The Al-5Si-1Cu-0.5Mg alloy ingot was melted at 750 °C and held for 10 min before the power of the electromagnetic stirrer set-up was turned off (as shown in Figure 2). When the temperature was decreased to 635 °C, the stirrer was started and the SiCp alloy was added. After that, the temperature of the furnace decreased at a constant cooling rate of 4 °C/min. When the temperature decreased to 585 °C, the stirrer was stopped. Then, the molten alloy was heated to 650 °C again and poured into a permanent mold. Subsequently, the fabricated 1.5 wt % SiCp/Al-5Si-1Cu-0.5Mg alloy was subjected to a solution treatment at 520 ± 2 °C for 6 h, and then quenched in room-temperature water. Those samples that had been solution-treated were aged at 175 ± 2 °C for up to 6 h and then air-cooled to room temperature. As a result, three types of samples were generated for the wear tests: the Al-5Si-1Cu-0.5Mg alloy, the 1.5 wt % SiCp/Al-5Si-1Cu-0.5Mg alloy, and the 1.5 wt % SiCp/Al-5Si-1Cu-0.5Mg alloy with T6.

Table 1. Chemical composition of the Al-5Si-1Cu-0.5Mg aluminum alloy used in the experiment (mass fraction, %).

Si	Cu	Mg	Ti	Fe	Al
5.13	1.27	0.55	0.16	0.13	Bal.

2.2. Wear Testing

The dry sliding wear test was carried out using a MMD-1 (Jinan Yihua Tribology Testing Technology Co., Ltd., Jinan, China) pin-on-disc apparatus at room temperature. The wear rates and friction coefficients results were the average value of the three tests. Pin samples were machined into rods of $\Phi 4.5$ mm $\times 11$ mm. The disc material was ASTM1045 steel of 45HRC. Before each test, the pin and disc surfaces were ground with 600, 1200, 1500, and 2000-grit SiC abrasive paper successively, polished, and then cleaned with ethanol. The surface of the samples was polished to a roughness less than 0.1 μm before wear testing. Three sliding speeds 100 r/min (0.188 m/s), 200 r/min (0.377 m/s), 300 r/min (0.565 m/s) and a sliding duration of 15 min were selected as metrics for the tests, while four different applied loads (15, 30, 60 and 90 N) were used. The weight of the pins was measured using a FA2204B electronic balance with an accuracy of ± 0.1 mg. All volume loss values were calculated by weight loss and the measured densities. Wear rates were estimated by dividing the volumetric wear loss by the sliding distance. Friction coefficients were the average values of the kinetic friction coefficients over a steady period of wear (after sliding for 5 min).

2.3. Characterization

A Nikon Eclipse MA200 (Nikon Metrology, Inc., Brighton, UK) optical microscope (OM) was used to observe and analyze the microstructure evolution of the samples. The samples that were used for the OM examination were mounted and then polished and etched in an aqueous solution composed of 0.5 vol % HF. The micro-hardness was measured using a HVS-1000A Vickers hardness instrument (Laizhou Huayin Testing Instrument Co., Ltd., Laizhou, China). The set load was 300 g and 10 s duration. To evaluate tensile properties, the test samples were machined into tensile test bars with a diameter of 6 mm and a gauge length of 30 mm, according to the ASTM B557M specification, which was carried out at room temperature using a SUNC UM5105 electro-servo testing machine with a tensile speed of 1 mm/min. To ensure the repeatability and consistency of the measurements, five samples were tested for tension under the same test conditon. The microstructures of the samples were observed using scan electron microscopy (FEI Quanta200F (FEI Trading (Shanghai) Co., Ltd., Shanghai, China)). EDS analysis was performed by an NCA 250XMax 50 instrument (Oxford instrument, Oxford, UK). The transmission electron microscopy (TEM) sample was cut from the composite with T6, then ground to less than 80 μm, and cut into a 3 mm diameter disk. Thin foil for TEM observation were prepared by twin-jet polishing with an electrolyte solution consisting of 10% HNO3 and 90% methanol below -30 °C. TEM observation was carried out by JEM-2100 (Japan Electronics Co., Ltd., Tokyo, Japan) microscope. The porosity of each composite was calculated by dividing the difference value between experimental and theoretical densities by theoretical densitie.

3. Results and Discussion

3.1. Microstructure Evolution

Figure 3a shows the microstucture of the Al-5Si-1Cu-0.5Mg alloy as cast-fabricated while using the electromagnetic stirring method. The base alloy microstructre consists of an α-Al (white) matrix surrounded by the eutectic Si (grey) phase of the needle or lamellar shape and the Fe-rich phase (γ phase) (sepia) of the intensive bone shape. The coarse grains and their inhomogeneous distribution produce stress concentration during loading and deformation, thus reducing the mechnical properties of the cast alloy. Figure 3b shows that the appropriate addition of SiCp (1.5 wt %) can refine the primary α-Al and the eutectic Si and improve their distribution. Under the environment of the semi-solid slurry, SiCp inhibits the diffusion of solute elements, such as Si, Fe, and other elements, thereby limiting the nucleation and growth of second phase [18]. On the other hand, SiCp provides a nucleation point for Si eutectic phase. Thus, the nucleation rate increases and the size of the phase reduces. Moreover, the aspect ratio of the eutectic Si phase can also be reduced due to the addition of a proper amount of SiCp [19]. The Fe-rich phase was also changed from an intensive bone shape into a dispersion type. Figure 3c illustrates the microstructure of SiCp/Al-5Si-1Cu-0.5Mg composites with heat-treatment. Spheroidization of the eutectic Si occurred during the solution treatment. After aging, there were some light grey γ phases precipitated. Table 2 shows the results of the mechanical tests of the samples. It could be seen that the addition of SiCp effectively improved the mechanical properties of the cast Al-5Si-1Cu-0.5Mg. Meanwhile, as Table 2 indicating, the mechnical properties of the SiCp/Al-5Si-1Cu-0.5Mg composites treated with T6 were the best of all the tested alloys. After T6 treatment, the hardness, ultimate strength, and elongation of the composite were 101.86 HV, 273.64 MPa, and 6.12%, respectively. When compared to the ones of the samples not treated with T6, these values were increased by 17%, 14.68%, and 17.02%, respectively.

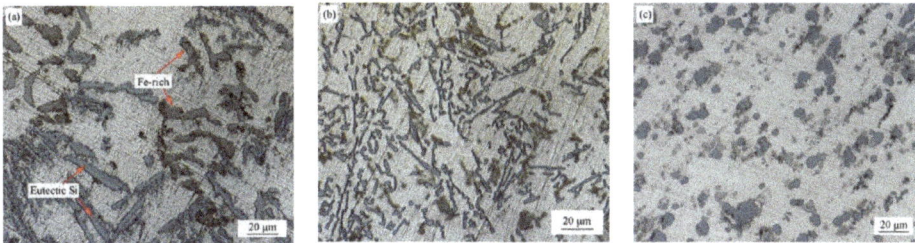

Figure 3. Microstructures of (**a**) Al-5Si-1Cu-0.5Mg alloy as cast, (**b**) SiCp/Al-5Si-1Cu-0.5Mg composites, and (**c**) T6 heat-treated SiCp/Al-5Si-1Cu-0.5Mg composites.

Table 2. Some properties of the samples used in the wear tests.

Samples	$\rho/(\text{g}\cdot\text{cm}^{-3})$	Porosity/%	Microhardness/HV	UTS/MPa
Al-5Si-1Cu-0.5Mg	2. 695	0.55	80.17 ± 3.50	191.56 ± 4.77
Al-5Si-1Cu-0.5Mg + SiCp	2. 699	0.61	87.06 ± 4.39	238.62 ± 5.19
Al-5Si-1Cu-0.5Mg + SiCp + T6	2. 698	0.65	101.86 ± 1.59	273.64 ± 6.47

Figure 4a is the TEM morphology of SiCp/Al-5Si-1Cu-0.5Mg composite after aging treatment. There are some needle-like precipitate phases (as the arrows shown in Figure 4a) with length of about 100 nm around the dislocation. Furthermore, EDS analysis (as shown in Figure 4b) of the precipitate phase revealed that aluminum and copper were the main compositions of the phase. According to Sjölander, E et al. [20], the precipitation phase could be Al_2Cu. The Al_2Cu phase contribute to improve the hardness of the alloy. Moreover, SiCp pinning on the boundaries that could stabilize the sub-microstructure and speed up the aging reaction and increase the rate of work hardening [21].

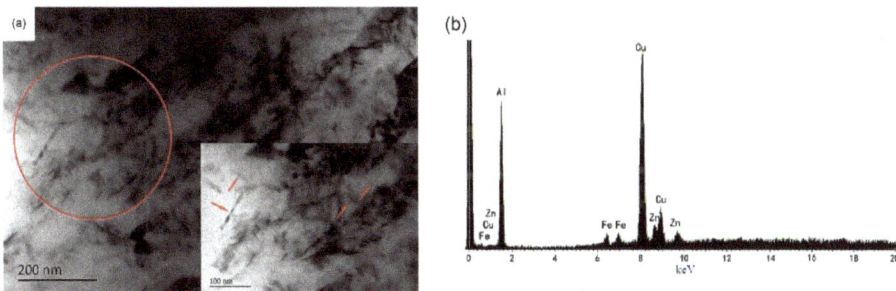

Figure 4. (**a**) TEM image and (**b**) EDS (energy dispersive spectroscopy) pattern of precipitates of SiCp/Al-5Si-1Cu-0.5Mg composite after aging treatment.

3.2. Wear Rates

As shown in Figure 5, the wear rates for the three alloys are plotted against applied loads. The wear rates increased with load. The wear rate of the as-cast Al-5Si-1Cu-0.5Mg alloy was the largest among the ones of the tested alloys. At low load (15 N), the wear rate of the as-cast SiCp/Al-5Si-1Cu-0.5Mg showed almost the same as the one of the base alloy. However, when the load upped to 90 N, the wear rate of the as-cast SiCp/Al-5Si-1Cu-0.5Mg was only about half of the one of the matrix alloy. The results indicated that the addition of SiCp significantly improved the wear resistance of the Al-5Si-1Cu-0.5Mg alloy. Especially, treated by T6, the SiCp/Al-5Si-1Cu-0.5Mg alloy achieved further improvement of the wear resistance. The wear rates of the three alloys were inversely proportional to the micro-hardness of the alloys, which was consistent with the Achard's law [22]. It has been reported that the wear behavior of the alloys was affected by their microstructure and mechanical properties [23]. As mentioned before, the as-cast Al-5Si-1Cu-0.5Mg alloy microstructre consisted of an α-Al matrix that was surrounded by the eutectic Si phase of the needle or lamellar shape, and the γ phase of the intensive bone shape. This configuration produced stress concentration on the boundary of the α-Al matrix, and the fracture occurred on the boundary interface, which caused the wear debris to fall out. With appropriate addition of SiCp, the shape of eutectic silicon was changed to a fine acicular or short stick or even a granular-like shape, and the γ phase was also distributed uniformly. Therefore, the interface bonding strength of the secondary phase with the matrix was reinforced, the mechanical and wear resistance property was improved, and the wear rate was decreased. Treated by T6, the eutectic silicon was spheroidal in shape, and some light grey γ phases were precipitated. As a result, the wear resistance property was improved.

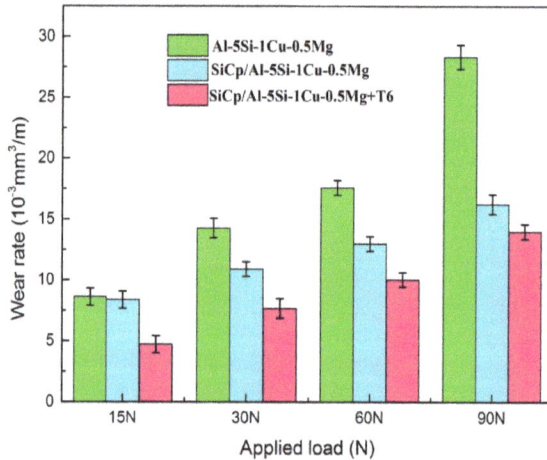

Figure 5. Wear rate values for samples under various applied loads.

Figure 6 illustrates that the wear rate decreases with the sliding speed increasing. As the sliding velocity increased, the worn surface became smooth and the amount of delaminating groove was decreased (Figure 7). At a higher sliding velocity, more frictional heat was generated. The heat generation could increase the plastic dedormation on the worn surface [23]. The deformation improved the hardening of the alloys. Therefore, the wear resistance property was improved.

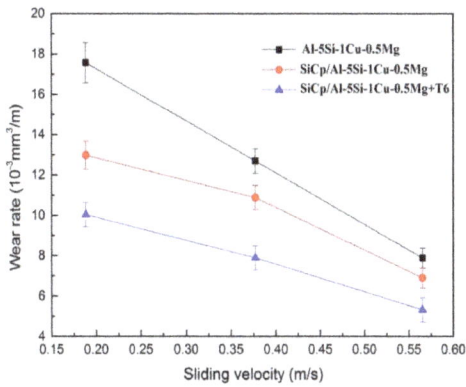

Figure 6. Wear rate values for samples at various Sliding velocity under a normal load of 60 N.

Figure 7. Worn surfaces of different alloys under a normal load of 60 N at different sliding speeds; (**a**) Al-5Si-1Cu-0.5Mg, 0.377 m/s, (**b**) Al-5Si-1Cu-0.5Mg, 0.565 m/s, (**c**) SiCp/Al-5Si-1Cu-0.5Mg, 0.377 m/s, (**d**) SiCp/Al-5Si-1Cu-0.5Mg, 0.565 m/s, (**e**) SiCp/Al-5Si-1Cu-0.5Mg with T6, 0.377 m/s, (**f**) SiCp/Al-5Si-1Cu-0.5Mg with T6, 0.565 m/s.

It has been recognized that under the condition of sliding with loading, strain hardening will be developed. The higher the sliding speed, the more pronounced the hardening of the alloy. SiCp particles and Al_2Cu phase could act as effective barriers to dislocation motion during subsurface deformation, which increased the rate of work-hardening. It can be considered as the main mechanism of the wear behavior improvement. So, when added with SiCp, Al-5Si-1Cu-0.5Mg alloy achieved a lower wear rate when compared with the one of the base alloy, and after T6 treatment, the wear rate value decreased further.

3.3. Friction Coefficient

The relationships between the frictional coefficients and applied loads are shown in Figure 8. The friction coefficient of the alloy treated with T6 was minimal under the same test conditions. This value decreased to its minimum (0.3255) at the lowest load (15 N). The collected friction coefficient values of the tested alloys under different applied loads are listed in Table 3.

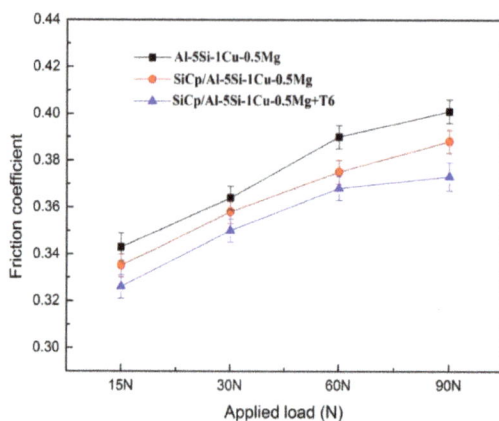

Figure 8. Coefficient of friction (COF) values for samples under various applied loads.

Table 3. Detailed Coefficient of friction (COF) values for samples at various applied loads.

Samples	15 N	30 N	60 N	90 N
Al-5Si-1Cu-0.5Mg	0.343 ± 0.003	0.364 ± 0.0025	0.390 ± 0.0025	0.401 ± 0.0025
Al-5Si-1Cu-0.5Mg + SiCp	0.335 ± 0.0025	0.358 ± 0.0025	0.375 ± 0.0025	0.388 ± 0.0025
Al-5Si-1Cu-0.5Mg + SiCp + T6	0.326 ± 0.0025	0.350 ± 0.0025	0.368 ± 0.0025	0. 373 ± 0.003

Bowden and Tabor's model for friction [24] regards friction as the resistance of asperities on one surface riding over the asperities of the matching surface. Under the same applied load, the friction coefficient and wear volume of the as-cast Al-5Si-1Cu-0.5Mg were all at maximum levels. The result showed that the wear debris fell out significantly between the asperities of the matching surface. This debris prevented the contact surfaces from sliding smoothly, and it significantly increased the friction force. The wear volume of the alloy with T6 treatment was minimal, few wear debris fell out, and the friction coefficient was low. In general, there is a proportional relationship among the friction coefficient and the microstructure or mechanical properties of the material [23,25].

The friction coefficient values increased with the applied load. When the two surfaces interacted, contact does not take place over the entire surface area. The real contact area consists of very small contact points at asperity tips, which are called micro-contacts [26]. When load was applied on the sample, the micro-contacts needed to have enough plastic deformation to bear the applied normal load. The higher the load, the greater the extent of the plastic deformation, which led to larger tribo-surface removal. Consequently, the contact area between the friction pairs increased as the load increasing, as well as friction. From the view of Bowden and Tabor's model for friction, at high loads, wear debris increased and became lodged between the asperities that resisted the motion at the interface. As a result, the friction force and friction coefficient increased.

Figure 9a illustrates that the friction coefficient values decreasing with the sliding speed increasing. As Figure 7b,d,f shows, smooth worn surface were observed on the samples without oxide layer. The smooth worn surface could be ascribed to the softening of the alloy at higher surface temperature during sliding at a high speed. The smooth worn surface led to a low friction coefficient value at a high sliding speed. Meanwhile, the fluctuation amplitude of the friction coefficient was reduced, as shown in Figure 9b–d.

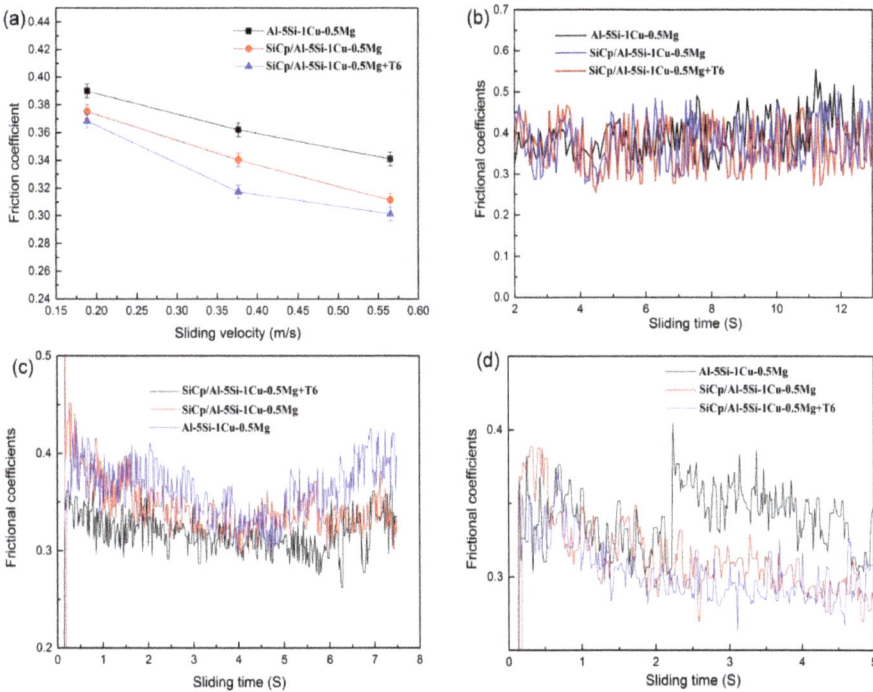

Figure 9. Friction coefficient for samples at various Sliding velocity under a normal load of 60 N; (**a**) the average friction coefficient, (**b**) the friction coefficient in time, 0.188 m/s, (**c**) the friction coefficient in time, 0.377 m/s, (**d**) the friction coefficient in time, 0.565 m/s.

3.4. Wear Mechanisms

In this study, SEM morphologies of 90 N load worn surfaces were captured, as shown in Figure 10. Figure 10b,d,f show the enlarged SEM micrograph of the circle area of the dotted line referred in Figure 10a,c,e, respectively. When combined with the energy spectrum analysis results (shown in Figure 11), the morphologies exhibited that there were five different wear operating mechanisms: abrasion, oxidation, delamination, plastic deformation, and adhesion.

Due to the easily oxidized nature of aluminum, frictional heating during sliding under high load that was caused the oxide film on the surface of the pin sample generated. Studies by Lashgari et al. [27] and Dwivedi et al. [28] showed that the existence of the oxide film improved the wear behavior and reduced the friction coefficient because of the lubrication effect. At the same time, the oxide film prevented metal-to-metal contact, resulting in a low wear volume. EDS analysis of areas A and B, marked in Figure 10c,d, were revealed in Figure 11a,b. By contrast, more Fe was clearly present in the A area. The white powder (Figure 10c,d) represented the Fe-rich wear debris, and also consisted of some other elements, such as: O, Al, Si, and Cu. The wear debris was closely related to abrasive wear.

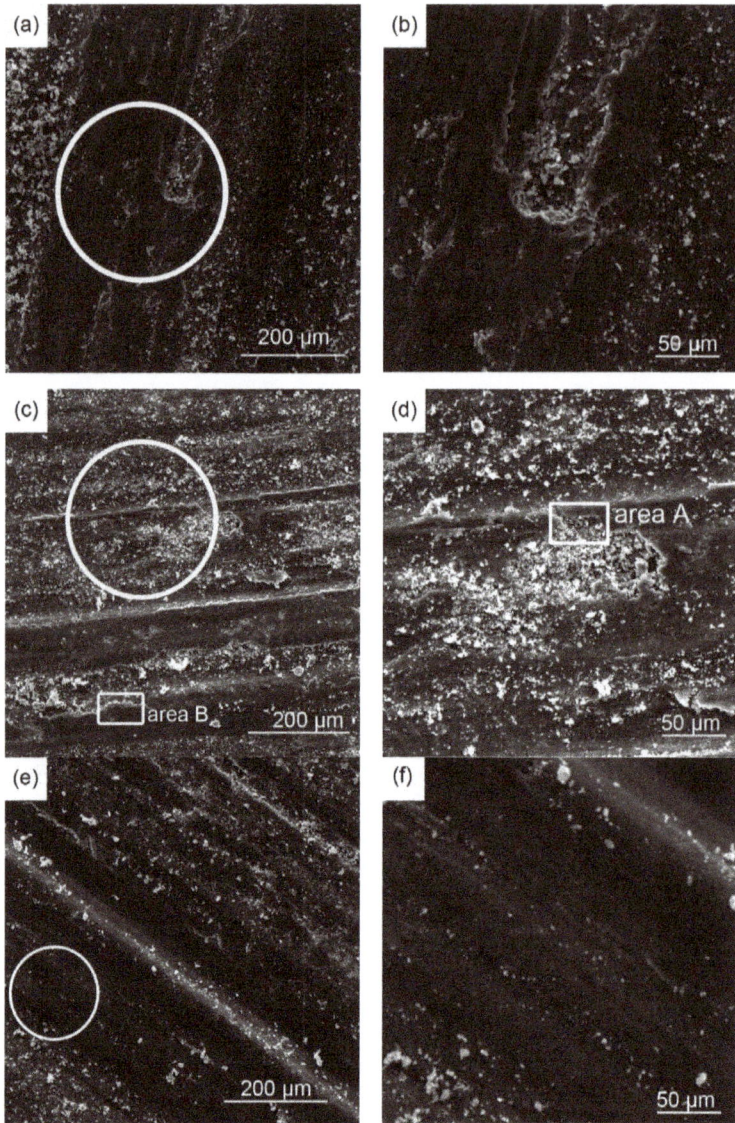

Figure 10. SEM images of samples under a normal load of 90 N at sliding speed of 0.188 m/s. (**a,b**) Al-5Si-1Cu-0.5Mg alloy, (**c,d**) SiCp/Al-5Si-1Cu-0.5Mg composites before T6 treatment and (**e,f**) after T6 treatment.

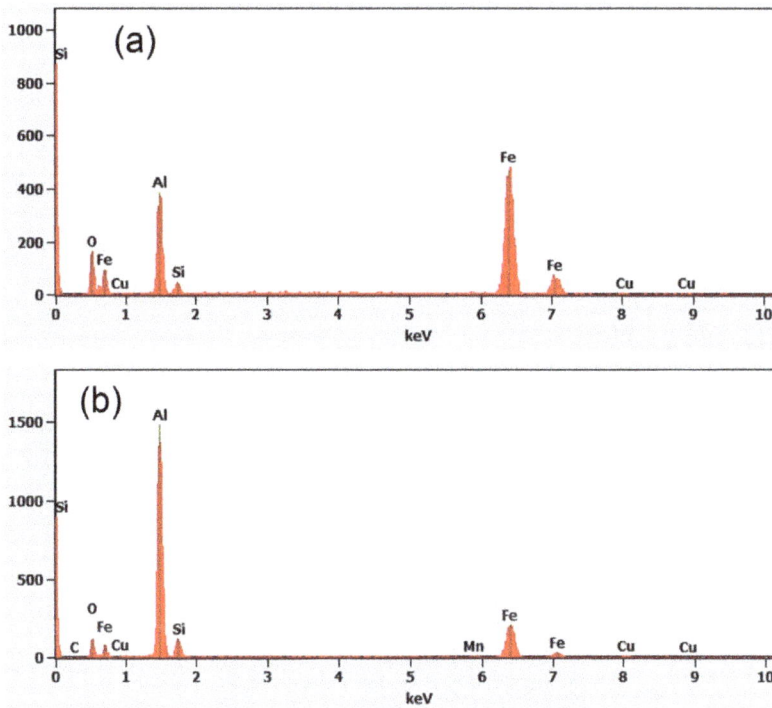

Figure 11. EDS maps of (**a**) area A and (**b**) area B.

Because of its fragile nature, the oxide film tended to be broken into fragments at high applied loads during the sliding process. The oxide fragments did not prevent metal-to-metal contact, and thus a high wear volume occurred. Wear behavior of the material was largely dependent on the microstructure of subsurface [29]. As Figure 10a,b shows, delaminating grooves can be distinguished, as well as plastic deformations around the grooves. Such a feature may often be linked with delamination wear, which is the result of cracks beneath the surface that grow and eventually join each other and eventually extend to the pin surface, leading to partial detachment of the wear layers [30,31]. As for the as-cast Al-5Si-1Cu-0.5Mg alloy, at the applied load of 90 N, a combination of delamination and plastic deformation was the dominant wear mechanism, and oxidation wear and abrasion wear also contributed to the wear.

As Figure 10c,d shows, many white powders (Fe-rich mechanical mixture), smooth flat surface (plastic deformation), and slight delamination wear were observed on the wear surface. These features indicated that oxidation wear and abrasion were the dominant wear mchanisms, and that plastic deformation and delamination also contributed to the wear. It can be conclued that SiCp improves the wear resistance of the matrix. Figure 10e,f show the SEM image of the SiCp/Al-5Si-1Cu-0.5Mg alloy with T6 treatment under a load of 90 N. Besides some tiny scratches, a large area of smooth grooves appear on the wear surface due to the plastic deformation. There are no cracks on the grooves. It indicates that the worn surface could guarantee smooth sliding after plastic deformation. When compared with the SiCp/Al-5Si-1Cu-0.5Mg alloy without T6 treatment, the wear resistance was significantly improved. The wear mechanisms were plastic deformation and abrasion wear.

As shown in Figure 7a,c,e, oxides, plastic deformation and grooves were on the worn surface of alloys at a sliding speed of 0.377 m/s and the load of 60 N. These features indicated that oxidation wear and abrasion were the dominant wear mchanisms, and that plastic deformation and delamination also contributed to the wear. Figure 7b shows that plastic deformation and scratches were formed on the worn surface of the as-cast Al-5Si-1Cu-0.5Mg alloy at a sliding speed of 0.565 m/s and the load of 60 N, which could suggest the abrasion wear mechanism. The same wear mechanism was also observed on the worn surfaces of the SiCp/Al-5Si-1Cu-0.5Mg and the SiCp/Al-5Si-1Cu-0.5Mg alloy with T6 treatment at the same sliding conditions. The plastic deformation of the as-cast Al-5Si-1Cu-0.5Mg alloy was the most seriously and the oxides were the minimum, owing to the high plastic deformation in contact zone and no oxides on the surface. Moreover, the existence of SiCp particles and Al_2Cu phase could help to protect the worn surfaces at a sliding of 0.565 m/s. The dominant wear mechanism of the SiCp/Al-5Si-1Cu-0.5Mg alloys with and without T6 treatment was plastic deformation, oxidation and abrasion wear at sliding speed of 0.565 m/s and load of 60 N (as Figure 7d,f shown).

Figure 12 shows the SEM micrographs of the debris of the SiCp/Al-5Si-1Cu-0.5Mg alloy with T6 treatment and the EDS analysis of point C. As shown in Table 4, O and Fe content were 10.48 and 33.88 wt %, respectively. The bulk wear debris was generated by the development of plastic strain and crack propagation in the subsurface of the composite. It indicated that delamination wear also occurred during the sliding process except for the wear mechanisms mentioned.

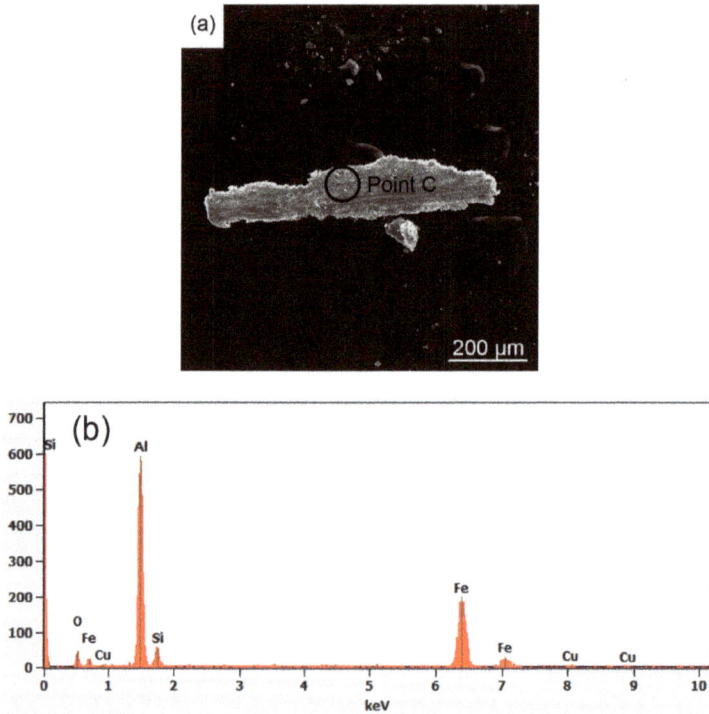

Figure 12. (a) Plate-like debris of the SiCp/Al-5Si-1Cu-0.5Mg sample after T6 heat treatment and (b) EDS result of point C.

Table 4. EDS results of area A, area B and point C (wt %).

Position	C-K	O-K	Al-K	Si-K	Mn-K	Fe-K	Cu-K
Area A	-	18.77	25.50	3.19	-	51.84	0.70
Area B	0.85	17.15	51.14	8.50	0.77	19.78	1.82
Point C	-	10.48	46.57	7.37	-	33.88	1.69

With the increase of applied loads, a large plastic strain appeared on the wear surface, and the wear rate increased (as Figure 5 shown). Figure 13a–c show the SEM images of the SiCp/Al-5Si-1Cu-0.5Mg composite with T6 samples under different loads. Numerous narrow scratches paralleling the sliding direction and few white powders can be distinguished on the wear suface at a load of 15 N (as shown in Figure 13a). Thus, abrasion wear was the dominant wear mechanism, and oxidation also caused some wear. When compared with Figure 13a, more white powder and deeper and wider scratches were shown in Figure 13b. This indicated that the wear intensity increased. Oxidation wear and abrasion were considered to be the dominant wear mechanisms. Some bulk wear debris were observed and a rugged flat surface appeared indistinctly on the wear surface (as shown in Figure 13c). This indicated that the plastic deformation was present on the wear surface. Meanwhile, the oxidation wear and abrasive wear was still observed. Figure 13d shows the SEM images of edging morphology of the as-cast Al-5Si-1Cu-0.5Mg pin sample at 90 N. The edging shape of the pin sample along the sliding direction was irregular due to the pressure of plastic deformation that was caused by the high applied load [32]. In other words, the wear resistance of the Al-5Si-1Cu-0.5Mg alloy was poor.

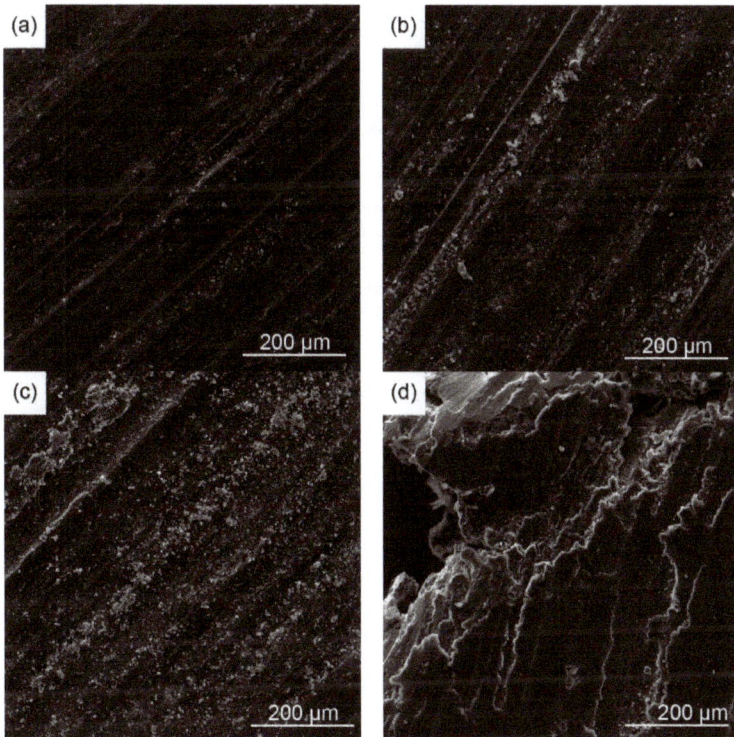

Figure 13. SEM images of SiCp/Al-5Si-1Cu-0.5Mg composites after T6 heat treatment under normal loads of (**a**) 15 N, (**b**) 30 N, (**c**) 60 N; (**d**) edging morphology of the as-cast Al-5Si-1Cu-0.5Mg pin sample at 90 N.

4. Conclusions

A pin-on-disc dry sliding wear test of three alloys processed by the electromagnetic stirring method was conducted at room temperature. The alloys included as-cast Al-5Si-1Cu-0.5Mg alloy, as-cast SiCp/Al-5Si-1Cu-0.5Mg, and SiCp/Al-5Si-1Cu-0.5Mg alloy with T6 heat treatment. The conclusions of this investigation are as follows:

(1) After T6 heat treatment, the morphology of the eutectic silicon was spheroidal, and some light grey γ phases precipitated. The mechnical properties of SiCp/Al-5Si-1Cu-0.5Mg alloy with T6 treatment performed the best of all. After T6 treatment, the hardness, ultimate strength, and elongation of the alloy were 101.86 HV, 273.64 MPa, and 6.12%, respectively. These values increased by 27%, 42.85% and 60.63%, respectively, when compared with the ones of as-cast Al-5Si-1Cu-0.5Mg.

(2) As the applied load increasing, the wear rates and friction coefficients of the alloy increased. The SiCp/Al-5Si-1Cu-0.5Mg alloy with T6 treatment achieved the lowest wear rate and lowest friction coefficient of all the alloys. This indicated that T6 heat treatment benefited for the improvement of the wear properties of the alloy.

(3) As the sliding speed increasing, the wear rates and friction coefficients of the alloy decreased. SiCp particles and Al_2Cu phase had positive effects on the wear behavior of the alloy. With the sliding speed increase, the amount of delaminating grooves on the worn surface was decreased and the worn surface became smooth. The SiCp/Al-5Si-1Cu-0.5Mg alloy with T6 treaement achieced the lowest wear rate and lowest friction coefficient of all the alloys.

(4) Abrasion and oxidation wear were dominant at low loads (15, 30 N). When the applied load increased to 60 N, plastic deformation was also observed on the wear surface. At a high applied load (90 N), plastic deformation and delamination were the main mechanisms of wear for the as-cast Al-5Si-1Cu-0.5Mg alloy. With SiCp addition, the degree of delamination was decreased. After T6 heat treatment, the worn surface of SiCp/Al-5Si-1Cu-0.5Mg alloy became smooth and flat, and no cracks were observed on the wear surface. As a result, the wear resistance of the SiCp/Al-5Si-1Cu-0.5Mg alloy was improved.

Acknowledgments: This work was supported by the national natural science foundation of china (No. 51364035), Loading Program of science and Technology of College of Jiangxi Province (No. KJLD14003).

Author Contributions: Ning Li and Hong Yan conceived and designed the experiments; Ning Li and Zhi-Wei Wang performed the experiments; Ning Li, Hong Yan and Zhi-Wei Wang analyzed the data; Ning Li, Hong Yan and Zhi-Wei Wang contributed reagents/materials/analysis tools; Ning Li, Hong Yan and Zhi-Wei Wang wrote the paper.

Conflicts of Interest: The author declares no conflict of interest.

References

1. Choi, S.W.; Cho, H.S.; Kang, C.S.; Kumai, S. Precipitation dependence of thermal properties for Al-Si-Mg-Cu-(Ti) alloy with various heat treatment. *J. Alloys Compd.* **2015**, *647*, 1091–1097. [CrossRef]
2. Rao, Y.S.; Yan, H.; Hu, Z. Modification of eutectic silicon and β-Al5FeSi phases in as-cast ADC12 alloys by using samarium addition. *J. Rare Earth.* **2013**, *9*, 916–922. [CrossRef]
3. Liu, G.; Li, G.D.; Cai, A.H.; Chen, Z.K. The influence of Strontium addition on wear properties of Al-20 wt % Si alloys under dry reciprocating sliding condition. *Mater. Des.* **2011**, *32*, 121–126. [CrossRef]
4. Jian, X.S.; Wang, N.J.; Zhu, D.G. Friction and wear properties of in-situ synthesized Al_2O_3 reinforced aluminum composites. *Trans. Nonferr. Met. Soc. China* **2014**, *24*, 2352–2358. [CrossRef]
5. Du, X.F.; Gao, T.; Liu, G.L.; Liu, X.F. In situ synthesizing SiC particles and its strengthening effect on an Al-Si-Cu-Ni-Mg piston alloy. *J. Alloys Compd.* **2017**, *695*, 1–8. [CrossRef]
6. Rajeev, V.R.; Dwivedi, D.K.; Jain, S.C. Effect of load and reciprocating velocity on the transition from mild to severe wear behavior of Al-Si-SiCp composites in reciprocating conditions. *Mater. Des.* **2010**, *31*, 4951–4959. [CrossRef]
7. Khosravi, H.; Akhlaghi, F. Comparison of microstructure and wear resistance of A356-SiCp composites processed via compocasting and vibration cooling slope. *Trans. Nonferr. Met. Soc. China* **2015**, *25*, 2490–2498. [CrossRef]
8. Gupta, A.K.; Prasad, B.K.; Pajnoo, R.K.; Das, S. Effects of T6 heat treatment on mechanical, abrasive and erosive-corrosive wear properties of eutectic Al-Si alloy. *Trans. Nonferr. Met. Soc. China* **2012**, *22*, 1041–1050. [CrossRef]
9. Singh, J.; Chauhan, A. Overview of wear performance of aluminum matrix composites reinforced with ceramic materials under the influence of controllable variables. *Ceram. Int.* **2016**, *42*, 56–81. [CrossRef]

10. Choi, J.P.; Hur, Y.M.; Nam, T.W.; Yoon, E.P. The effect of frequency of electromagnetic vibration on the micro structure in hypoeutectic Al-Si Alloy. *Solid State Phenom.* **2006**, *116–117*, 213–216. [CrossRef]

11. Yu, H.Q.; Zhu, M.Y. Effect of electromagnetic stirring in mold on the macroscopic quality of high carbon steel billet. *Acta Metall. Sin.* **2009**, *22*, 461–467. [CrossRef]

12. Zhu, G.L.; Xu, J.; Zhang, Z.F.; Bai, Y.L.; Shi, L.K. Annular electromagnetic stirring—A new method for the production of semi-solid A357 aluminum alloy slurry. *Acta Metall. Sin.* **2009**, *22*, 408–414. [CrossRef]

13. Metan, V.; Eigenfeld, K.; Rabiger, D.; Leonhardt, M.; Eckert, S. Grain size control in Al-Si alloys by grain refinement and electronmagnetic stirring. *J. Alloys Compd.* **2009**, *487*, 163–172. [CrossRef]

14. Liu, Z.; Mao, W.M.; Liu, X.M. Characterization on morphology evolution of primary phase in semisolid A356 under slightly electromagnetic stirring. *Trans. Nonferr. Met. Soc. China* **2010**, *20*, 805–810. [CrossRef]

15. Dwivedi, S.P.; Sharma, S.; Mishra, R.K. Microstructure and mechanical properties of A356/SiC composites fabricated by electromagnetic stir casting. *Procedia Mater. Sci.* **2014**, *6*, 1524–1532. [CrossRef]

16. Liu, G.P.; Wang, Q.D.; Liu, T.; Ye, B.; Jiang, H.Y.; Ding, W.J. Effect of T6 heat treatment on microstructure and mechanical property of 6101/A356 bimetal fabricated by squeeze casting. *Mater. Sci. Eng. A* **2017**, *696*, 208–215. [CrossRef]

17. Yan, H.; Wan, J.; Nie, Q. Wear Behavior of Extruded Nano-SiCp Reinforced AZ61 Magnesium Matrix Composites. *Adv. Mech. Eng.* **2015**, *5*, 489528. [CrossRef]

18. Rohatgi, P.K.; Ray, S.; Asthana, R.; Narendranath, C.S. Interfaces in cast metal-matrix composites. *Mater. Sci. Eng. A* **1993**, *162*, 163–174. [CrossRef]

19. Wang, W.; Ajersch, F.; Löfvander, J.P.A. Si phase nucleation on SiC particulate reinforcement in hypereutectic Al-Si alloy matrix. *Mater. Sci. Eng. A* **1994**, *187*, 65–75. [CrossRef]

20. Sjölander, E.; Seifeddine, S. The heat treatment of Al-Si-Cu-Mg casting alloys. *J. Mater. Process. Technol.* **2010**, *210*, 1249–1259. [CrossRef]

21. Chu, H.S.; Liu, K.S.; Yeh, J.W. Study of 6061-Al2O3p composites produced by reciprocasting extrusion. *Metall. Mater. Trans. A* **2000**, *31*, 2587–2596. [CrossRef]

22. Archard, J.F. Contact and rubbing of flat surfaces. *J. Appl. Phys.* **1953**, *24*, 981–988. [CrossRef]

23. Yan, H.; Wang, Z.W. Effect of heat treatment on wear properties of extruded AZ91 alloy treated with yttrium. *J. Rare Earths* **2016**, *34*, 308–314. [CrossRef]

24. Bowden, F.P.; Tabor, D. *The Friction and Lubrication of Solids*; Clarendon Press: London, UK, 1964; pp. 1–8. ISBN 9780198507772.

25. An, J.; Li, R.G.; Lu, Y.; Chen, C.M.; Xu, Y.; Chen, X.; Wang, L.M. Dry sliding wear behavior of magnesium alloys. *Wear* **2008**, *265*, 97–104. [CrossRef]

26. Ramezani, M.; Ripin, Z.M. A friction model for dry contacts during metal-forming process. *Int. J. Adv. Manuf. Technol.* **2010**, *51*, 93–102. [CrossRef]

27. Lashgari, H.R.; Sufizadeh, A.R.; Emamy, M. The effect of strontium on the microstructure and wear properties of A356–10%B4C cast composites. *Mater. Des.* **2010**, *31*, 2187–2195. [CrossRef]

28. Dwivedi, D.K.; Arjun, T.S.; Thakur, P.; Vaidya, H.; Singh, K. Sliding wear and friction behaviour of Al–18% Si–0.5% Mg alloy. *J. Mater. Process. Technol.* **2004**, *152*, 323–328. [CrossRef]

29. Abouei, V.; Shabestari, S.G.; Saghafian, H. Dry sliding wear behaviour of hypereutectic Al–Si piston alloys containing iron-rich intermetallics. *Mater. Charact.* **2010**, *61*, 1089–1096. [CrossRef]

30. Suh, N.P. An overview of the delamination theory of wear. *Wear* **1977**, *44*, 1–16. [CrossRef]

31. Zhu, J.B.; Yan, H. Fabrication of an A356/fly-ash-mullite interpenetrating composite and its wear properties. *Ceram. Int.* **2017**, *43*, 12996–13003. [CrossRef]

32. Asl, K.M.; Masoudi, A.; Khomamizadeh, F. The effect of different rare earth elements content on microstructure, mechanical and wear behavior of Mg–Al–Zn alloy. *Mater. Sci. Eng. A* **2010**, *527*, 2027–2035.

applied
sciences

MDPI

Article

Impact of Alloying on Stacking Fault Energies in γ-TiAl

Phillip Dumitraschkewitz [†], Helmut Clemens, Svea Mayer and David Holec *

Chair of Physical Metallurgy and Metallic Materials, Department of Physical Metallurgy and Materials Testing, Montanuniversität Leoben, Leoben A-8700, Austria; phillip.dumitraschkewitz@unileoben.ac.at (P.D.); helmut.clemens@unileoben.ac.at (H.C.); svea.mayer@unileoben.ac.at (S.M.)
* Correspondence: david.holec@unileoben.ac.at; Tel.: +43-3842-402-4211
† Current address: Chair of Nonferrous Metallurgy, Department of Metallurgy, Montanuniversität Leoben, Franz-Josef-Str. 18, Leoben 8700, Austria

Received: 21 October 2017; Accepted: 15 November 2017 ; Published: 21 November 2017

Abstract: Microstructure and mechanical properties are key parameters influencing the performance of structural multi-phase alloys such as those based on intermetallic TiAl compounds. There, the main constituent, a γ-TiAl phase, is derived from a face-centered cubic structure. Consequently, the dissociation of dislocations and generation of stacking faults (SFs) are important factors contributing to the overall deformation behavior, as well as mechanical properties, such as tensile/creep strength and, most importantly, fracture elongation below the brittle-to-ductile transition temperature. In this work, SFs on the {111} plane in γ-TiAl are revisited by means of ab initio calculations, finding their energies in agreement with previous reports. Subsequently, stacking fault energies are evaluated for eight ternary additions, namely group IVB–VIB elements, together with Ti off-stoichiometry. It is found that the energies of superlattice intrinsic SFs, anti-phase boundaries (APBs), as well as complex SFs decrease by 20–40% with respect to values in stoichiometric γ-TiAl once an alloying element X is present in the fault plane having thus a composition of Ti-50Al-12.5X. In addition, Mo, Ti and V stabilize the APB on the (111) plane, which is intrinsically unstable at 0 K in stoichiometric γ-TiAl.

Keywords: titanium aluminides; stacking fault energies; density functional theory

---

1. Introduction

Titanium aluminides are intermetallic compounds and alloys with a wide reach for high-temperature applications. These range from low-pressure turbine blades in the aircraft industry to turbocharger turbine wheels and valves in the automotive industry [1–3]. Their outstanding properties include low mass density, high specific strength and stiffness and good creep properties up to 750 °C. They outperform titanium alloys by their good oxidation behavior and burn resistance [2]. In contrast to ceramic materials, titanium aluminides also exhibit the ability to plastically deform at room temperature [4].

Current state-of-the-art TiAl alloys consist of a close to face-centered cubic (fcc) γ-TiAl phase, a hexagonal α_2-Ti$_3$Al phase, a body-centered cubic (bcc) β_0-TiAl phase, and, occasionally, additional minor phases [1]. However, the main constituent is the γ-TiAl phase, which fundamentally influences the alloy properties during processing, e.g., hot-forging and application.

The γ-TiAl phase has an L1$_0$ structure which is a tetragonally strained fcc lattice with (001) planes occupied alternatively by Ti and Al atoms. Therefore, stacking faults (SFs) become a topic of a great importance, similar to fcc metals, in which a stacking fault is a deviation from the normal stacking sequence . . . $ABCABC$ . . . of the (111) planes. The stacking fault energy (SFE), γ, an energy stored by the stacking fault defect, has a huge impact on a plastic deformation behavior of fcc metals. It determines the spreading of dissociated partial dislocations and, therefore, influences the cross-slip properties of screw dislocations. It is observed that metals with smaller stacking fault energies exhibit more mechanical twinning whereby they possess an additional deformation mechanism [5]. Especially, this effect was utilized in the development of novel high-strength γ-TiAl-based alloys exhibiting a certain ductility at room temperature. Here, the reader is referred to [1,2] and the papers cited therein.

Measuring the stacking fault energies experimentally is, nevertheless, non-trivial. Most commonly, they are measured indirectly using transmission electron microscopy (TEM) from the separation of partial dislocations. Such an experimental procedure is, however, time consuming and deals with all the difficulties related to a multi-phase complex alloy (non-homogeneous concentrations, stresses, sample preparation, to name a few). Consequently, experimental works dealing with this topic in γ-TiAl are scarce [6–9]. Hence, theoretical studies of stacking fault energies comprise a welcome alternative to examine stacking faults. These included both ab initio methods [10–14], as well as empirical atomistic modeling [15–17].

Nonetheless, the "real" alloys go beyond the simple binary Ti-Al system [1,2,4]. It is therefore of immense interest to know what impact alloying has on the SFEs and, consequently, on the expected deformation mechanisms. A rare example of such modeling effort is the study of Woodward and MacLaren [18], who used the coherent potential approximation (CPA) to investigate the impact of Nb and Cr on SFEs in γ-TiAl. This formalism, however, does not allow for any relaxations of the local atomic environments. Moreover, information on other alloying elements, namely the group IVB, VB, and VIB transition metal (TM) elements, commonly used experimentally, is still missing. The aim of the present study is to fill this gap by investigating the impact of ternary alloying elements on planar faults in γ-TiAl using ab initio techniques.

2. Methods

2.1. Geometry of Planar Defects in γ-TiAl

The stacking faults in fcc materials are irregularities of stacking the (111) planes. Let A, B and C denote the three configurations of the close-packed (111) planes, being mutually displaced by a vector $1/6[\bar{2}11]$ in coordinates related to the conventional cubic cell (containing four atoms). The perfect stacking $\ldots ABCABCABC\ldots$ could change to $\ldots ABC\underline{A}CBA\ldots$, where the $\underline{A}$ denotes the position of a twin boundary (a twin mirror plane). Other faults include a missing or an extra plane without a mirror. The former is called intrinsic stacking fault (ISF) and is described by a stacking sequence $\ldots ABCA.CABC\ldots$, while the latter is called extrinsic stacking fault (ESF) and corresponds to $\ldots ABCA\underline{C}BCABC\ldots$.

The situation is somewhat more complicated for the L1_0 structure as in the case of γ-TiAl. Its lattice is a slightly tetragonally-deformed fcc lattice ($c/a \approx 1.016$; hence, the [001] direction is not equivalent with [100] or [010] any more). Moreover, the (001) planes are alternatively occupied by Al and Ti atoms. Consequently, out of the three displacement vectors $\vec{b}_1 = 1/6\langle\bar{2}11\rangle$, $\vec{b}_2 = 1/6\langle1\bar{2}1\rangle$ and $\vec{b}_3 = 1/6\langle11\bar{2}\rangle$, being equivalent in the fcc structure (and all producing ISF), only $\vec{b}_3$ creates a fault-preserving local chemical neighborhood of the atoms. The resulting fault is called superlattice ISF (SISF). $\vec{b}_1$ and $\vec{b}_2$, in addition to the stacking fault, also alter the chemical occupation of the sites, yielding so-called complex stacking faults (CSFs). Finally, a displacement $\vec{b}_4 = \vec{b}_1 - \vec{b}_2 = 1/2\langle\bar{1}10\rangle$ results in an undistorted lattice, but altered lattice occupations, so-called anti-phase, and the corresponding fault plane is therefore called the anti-phase boundary (APB).

SFE expresses the energy difference between the faulted, E_{faulted}, and perfect, E_{perfect}, configurations per unit area, A:

$$\gamma = \frac{E_{\text{faulted}} - E_{\text{perfect}}}{A}. \tag{1}$$

This energy can be calculated not only for the special translations corresponding to SFs as described above, but for any displacement vector $\vec{b}$, hence yielding the generalized stacking fault energy (GSFE) surface [19]. In addition, to the actual values of SFEs, the GSFE surface contains also energy barriers that need to be overcome when an SF is created. In this work, we discuss SFs on the {111} planes; hence, GSFE surfaces are evaluated along u and v coordinates decomposing $\vec{b} = u\frac{1}{2}\langle \bar{1}10] + v\frac{1}{2}\langle 11\bar{2}]$. It is worth noting that generalization of ESF for the L1$_0$ structure yields the superlattice extrinsic stacking fault (SESF), which is, however, not compatible with a single shear plane. Since it is not contained by the simple GSFE surface, it will not be discussed any further in this work. For similar reasons, we do not include twin boundaries or APBs on the {010} planes.

2.2. Modeling of SFs

One possible approach for simulating SFs is to build a $1 \times 1 \times n$ supercell of cells having the fault planes perpendicular to the $\vec{a}_3$ lattice vector. Subsequently, a section of the supercell is displaced in the $a_1 a_2$ plane (fault plane) so as to produce an SF with $\vec{b}$, and if necessary, a layer is removed in order to restore the periodic boundary conditions (Figure 1). In some cases (e.g., APB), this approach implies that there are two SFs per supercell.

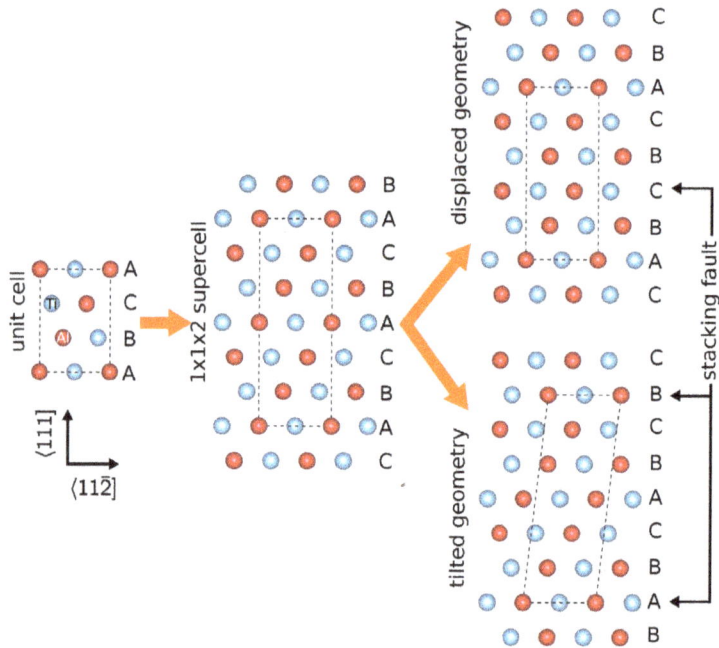

Figure 1. Schematic drawing demonstrating the two possible modeling approaches to SFs in the γ-TiAl with the L1$_0$ structure.

In this work, we assumed another approach, which allows for a straightforward evaluation of GSFE. The Cartesian positions of all atoms remain unchanged in the whole supercell, and instead, the supercell vector $n\vec{a}_3$ is tilted to become $n\vec{a}_3 + \vec{b}$. We note that, strictly speaking, these two approaches are inequivalent if the atomic relaxations are performed along the supercell lattice vector directions. In particular, to calculate SFE, one typically fixes the atomic positions in the SF plane and relaxes the positions (forces) in the $\vec{a}_3$ direction. This direction is perpendicular to the SF plane in the former approach with the displaced geometry, while it is slightly tilted away in the latter approach. The actual difference on the predicted results will be discussed later in the text and will be shown to be negligible in the present case provided the supercell size n is large enough (the tilt decreases with increasing n).

2.3. Computational Details

The present quantum mechanical calculations are based on density functional theory (DFT) [20,21] as implemented in the Vienna Ab initio Simulation Package (VASP) [22,23]. The basis set contained plane waves corresponding to energies lower than 400 eV. Our convergence tests showed that together with the 17 × 17 × 17 Monkhorst–Pack mesh sampling the Brillouin zone of the γ-TiAl conventional cell with 2 Ti and 2 Al atoms, this cut-off energy should guarantee total energy accuracy in the range of a few meV/at. The exchange and correlation (xc) effects were treated within the generalized gradient approximation as parametrized by Perdew and Wang (GGA-PW91) [24]. The structural relaxations were carried out until the forces changed less than 0.03 eV/Å, while each electronic loop was converged until the total energy changes were smaller than 10^{-7} eV (per simulation box). SFEs for the stoichiometric γ-TiAl were evaluated for $n \geq 5$ (at least 15 {111} planes), while the alloying studies, which required laterally larger cells, were performed for $n = 2$ (in total, six {111} planes).

3. Results and Discussion

3.1. SFE in γ-TiAl

The calculated SFEs are summarized in Table 1. In addition to the method based on tilting the supercell vectors, also a complementary method based on displacing a rigid block of the supercell [13,14] was employed. The differences are negligible. Similarly, our calculations predict only a small decrease of the SFE values (below 6%) when the local density approximation (LDA) is used instead of GGA. Test calculations with respect to the supercell size also suggest that the used models are large enough to yield converged results. The values calculated here lie in the range of previously published predictions based on DFT. Consequently, the GGA-PW91 xc potential together with the tilted supercell geometry were used for all subsequently discussed results.

Table 1. Calculated SF energies (in mJ/m^2) compared with available literature data for γ-TiAl. Differences in calculation methods are noted.

	APB	CSF	SISF	Note
present work	717	415	188	GGA-PW91, VASP, tilted supercells
	635	370	173	GGA-PW91, VASP, tilted supercells, fully relaxed
	711	414	188	GGA-PW91, VASP, displaced supercells
	694	392	179	LDA, VASP, tilted supercells
[10]	710	314	134	LDA, FP-LMTO, tilted supercells(?)
[11]	756	420	184	LDA, FP-LAPW, tilted supercells
[12]	499	329	137	GGA-PW91, CASTEP, tilted supercells
[13]		355	184	GGA-PW91, VASP, displaced supercells
[14]	663	400	170	GGA-PW91, VASP, displaced supercells

Figure 2 shows the calculated GSFE surface with its profiles along dissociation paths for $\frac{1}{2}\langle01\bar{1}]$ and $\frac{1}{2}\langle11\bar{2}]$ superdislocations [1]:

$$\langle01\bar{1}] \rightarrow \frac{1}{6}\langle11\bar{2}] + \text{SISF} + \frac{1}{6}\langle\bar{1}2\bar{1}] + \text{APB} + \frac{1}{2}\langle01\bar{1}] , \tag{2}$$

$$\frac{1}{2}\langle11\bar{2}] \rightarrow \frac{1}{6}\langle11\bar{2}] + \text{SISF} + \frac{1}{6}\langle2\bar{1}\bar{1}] + \text{APB} + \frac{1}{6}\langle11\bar{2}] + \text{CSF} + \frac{1}{6}\langle\bar{1}2\bar{1}] . \tag{3}$$

The APB turns out to be an unstable fault at 0 K as is does not correspond to a local minimum on the GSFE surface. This is in agreement with previous reports [10,14]. On the contrary, both SISF and CSF are stable; in order to produce them, however, barriers of ≈335 mJ/m^2 and ≈550 mJ/m^2, respectively, must be overcome when starting from a perfect stacking. We note, that these barriers correspond to the GSFE energy profiles along rigid pathways (as visualized in Figure 2), without attempting to estimate exact lowest-energy pathways. Even more importantly, it becomes apparent that the displacement vectors, $\vec{b}$, of the planar defects (i.e., positions of the local minima on the GSFE surface) do not correspond with the geometrically-determined ones based on the hard-sphere model. In particular, the CSF is shifted by approximately $-0.018\langle 11\bar{2}]$ ($0.11(\vec{b}_{APB} - \vec{b}_{CSF})$), which is clearly visible also in the profile in Figure 2a. This actually lowers the γ_{CSF} from $415\,\text{mJ/m}^2$ down to $383\,\text{mJ/m}^2$. Such an effect with similar magnitudes of displacements was previously predicted [10,11].

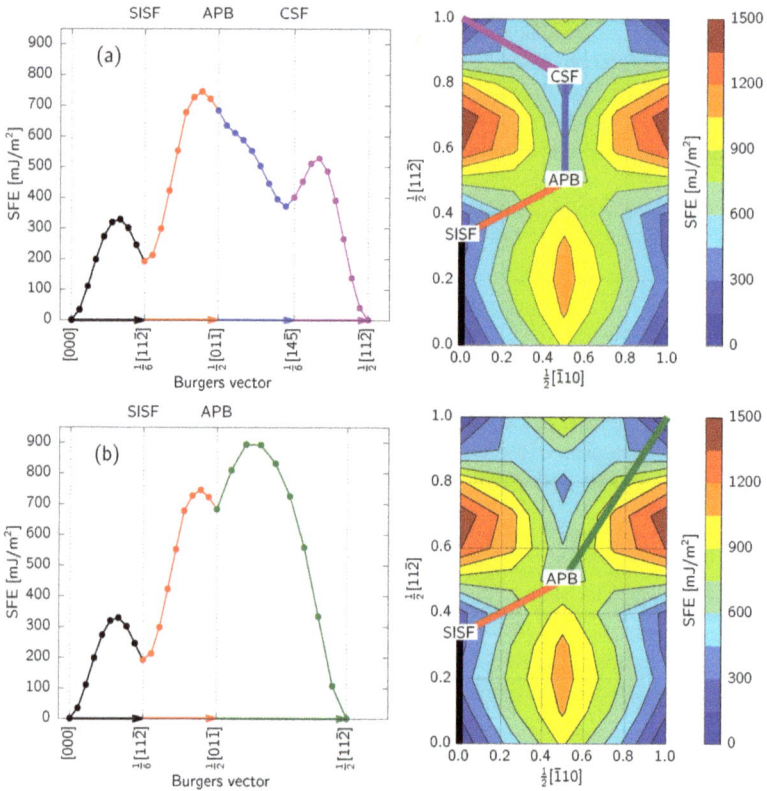

Figure 2. Energy profile along the dissociation path according to (**a**) Equation (2) and (**b**) Equation (3). The right panels show the dissociation path overlayed on the calculated {111} GSFE surface of the γ-TiAl lattice.

During the construction of the GSFE surface, the atomic coordinates in the {111} plane were fixed, while a relaxation along the $\langle 111]$ direction was allowed. Nonetheless, the SF energy can be decreased even more by fully relaxing the atomic positions once the geometry of the faulted material is trapped near the local energy minimum. This leads to a further decrease of SFEs to $\gamma_{APB} = 635\,\text{mJ/m}^2$, $\gamma_{CSF} = 370\,\text{mJ/m}^2$ and $\gamma_{SISF} = 173\,\text{mJ/m}^2$ (see Table 1).

3.2. Impact of Alloying Elements

In order to tune various application-related properties, alloying elements are introduced to γ-TiAl-based alloys. This section thus presents predictions of the impact of early transition metals (TMs, group IVB–VIB elements) on SFEs of γ-TiAl. Experimentally, the most relevant are Ti-rich compositions [1,3,25]. Moreover, the early TMs preferably occupy the Ti sublattice in γ-TiAl [26,27]. Therefore, a scenario with a Ti anti-site (Ti atom on the Al sublattice) and a TM substitutional atom on the Ti sublattice was considered. The supercells consisted of six {111} planes, each containing eight atoms (a 2×2 supercell laterally), and the calculations were performed using the tilted $\vec{a}_3$ geometry (Figure 1). Both the alloying element (having a concentration $1/48 \approx 0.02$) and the Ti anti-site were put in the fault plane. The results are summarized in Figure 3.

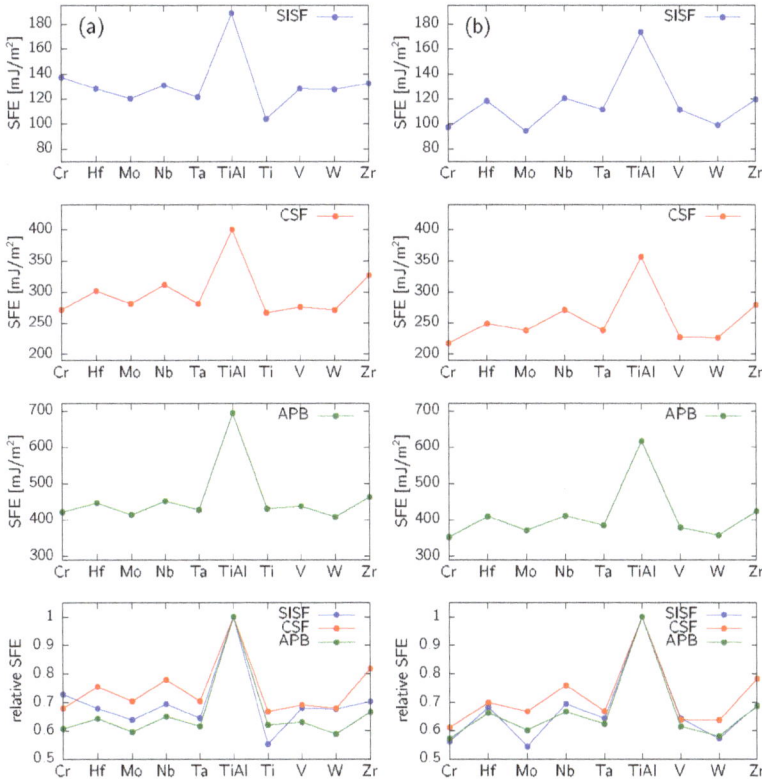

Figure 3. Impact of alloying on γ-TiAl with a composition Ti-48Al-2X (X = transition metal element): (**a**) SFs fixed to their geometrically-dictated configurations and relaxed only in the perpendicular direction and (**b**) atomic positions fully relaxed.

The SFEs are predicted to significantly decrease by 20–40% as a result of the alloying (see Figure 3a; atomic positions were relaxed only in the direction perpendicular to the fault plane). Regarding the SISF, the biggest impact has Ti simply leading to a bigger Ti/Al non-stoichiometry (Al/(Ti + Al) = 47.9) in comparison with the ternary systems (Al/(Ti + Al) = 48.9). From the ternary additions, the most pronounced effect has Mo (γ_{SISF} drops by 36%) closely followed by Ta. The largest reduction of the CSF energy is caused by Cr, V and W, while the APB energy is most significantly decreased by Mo and W. The relative reduction of SFEs is even more pronounced when the full atomic relaxation is performed (Figure 3b).

The SFEs presented in (Figure 3 contained the alloying element directly in the fault plane. In order to see how localized the alloying impact is, we calculated the SFEs for the off-stoichiometric TiAl, i.e., 52Ti-48Al, as a function of the distance of the substitutional Ti atom from the fault plane. It turns out that the huge SFE reduction happens only in the case when Ti is directly in the SF ($N = 0$ in

Figure 4). When positioned in the neighboring layer ($N = 1$), SFE is still decreased with respect to the stoichiometric value (most significantly in the SISF case). For $N \geq 2$, the SFEs stay practically constant, though slightly lower than in the stoichiometric case.

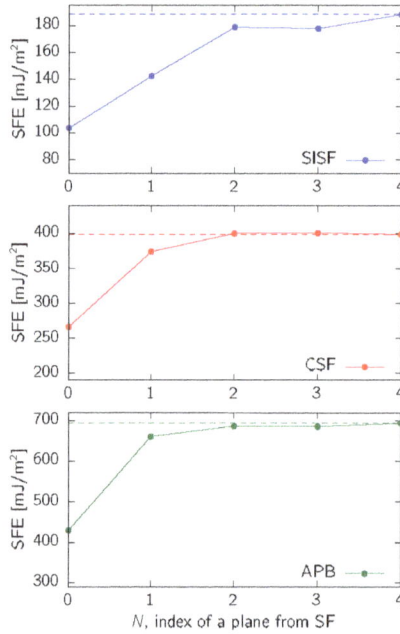

Figure 4. Dependence of SFEs on the location of the substitutional atom (here, Ti). N labels the layer number away from the fault plane (located at $N = 0$), and the dashed line represents a reference value for the stoichiometric TiAl (Table 1).

Alloying does not only influence the SFE, but the whole GSFE surface including its topology. Figure 5 gives an example of the GSFE profiles along the two dissociation paths, Equations (2) and (3), for γ-TiAl + Nb. When compared with analogous profiles in Figure 2, two observations can be made. Firstly, in addition to lower values of the local minima, also the local maxima (transformation barriers) are lower, leading to an overall easier creation of SFs. Secondly, while the APB was predicted to be unstable for stoichiometric γ-TiAl, a shallow local minimum is developed in the case of TiAl + Nb, suggesting that Nb stabilizes this fault geometry. A similar effect was predicted also for the V and Ti off-stoichiometry (52Ti-48Al).

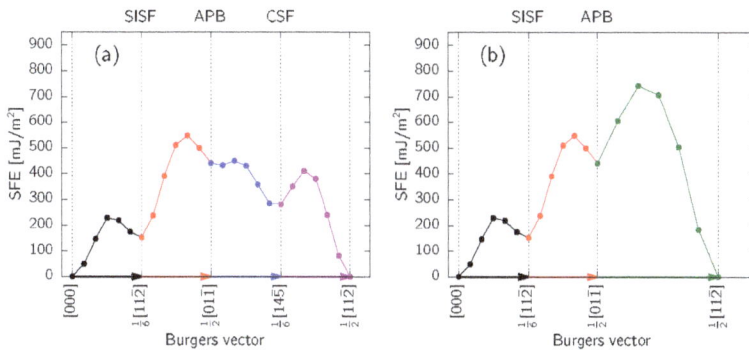

Figure 5. GSFE profile for Ti-48Al-2Nb along the (**a**) Equation (2) and (**b**) Equation (3) dissociation paths.

3.3. Comparison with Experiment

A critical point for validating theoretical predictions is a comparison with available experimental data. The reported values for SISF energies decrease with Ti content from $\approx$140 mJ/m^2 for Ti-54Al [6,7] to $\approx$97 mJ/m^2 for Ti-49.6Al [8] to 67 mJ/m^2 for Ti-48Al [8]. While these values are lower than those predicted by the ab initio methods (which, however, are all mutually consistent, cf. Table 1), the theory and the experiment agree on ordering the SFE as $\gamma_{APB} > \gamma_{CSF} > \gamma_{SISF}$ ([6,7] and Table 1), as well as on the decreasing value of SISF energy with increasing Ti content ([8] and Figures 3 and 4). Seemingly better agreement is obtained between experiment and semi-empirical atomistic simulations yielding SISF energies in the range from 3 to 250 mJ/m^2 (see [7,10] and the references therein). This is, however, not surprising since SFEs are often contained in the set of data used for fitting the interatomic potentials. Moreover, our own molecular dynamics calculations yielded an apparent disagreement between the SFEs' hierarchy using two embedded-atom method potentials [28]: while the one parametrized by Zope and Mishin [29] yields $\gamma_{CSF} > \gamma_{APB} > \gamma_{SISF}$, the one by Farkas and Jones [30] results in the same ordering as the ab initio calculations.

SISF energies for Nb-containing γ-TiAl were experimentally measured to be 63 mJ/m^2 for Ti-48Al-1Nb [9] and 66 mJ/m^2 for Ti-45Al-10Nb [8]. Using the value of $\gamma_{SISF} = 97$ mJ/m^2 reported for almost stoichiometric TiAl [8], Nb causes γ_{SISF} to drop by $\approx$30%, a value in an excellent agreement with our predictions (cf. Figure 3). It is worth noting that since we estimated that Nb impacts SFE only when present directly in the fault plane (or a plane next to it) and since each of the {111} planes in our supercells contained eight atoms, we predict a drop by $\approx$30% effectively for Ti-50Al-12.5Nb.

Consequently, we conclude that the ab initio predicted SFE values agree only semi-quantitatively with the experimental observations, but are expected to be representative of the prevailing trend: early TMs (group IVB–VIB elements) decrease SFE in Ti-rich γ-TiAl alloys. They increase valence electron concentration, which according to Thornton's semi-empirical observation, leads to a decrease of the SFE [31]. Therefore, alloying these elements in γ-TiAl is expected to lead to various effects, which can be utilized for advanced alloy design. For example, a decreased SFE is accompanied by a dissociation of dislocations and, hence, decreasing their mobility (climb rate) at elevated temperatures. On the other hand, a lower SFE enhances the propensity for mechanical twinning, thus increasing the ductility at temperatures below the brittle-to-ductile transition temperature. However, it should be noted that an alloying element, before being eventually selected, must fulfil also other criteria, e.g., influence on oxidation behavior, etc.

4. Conclusions

In this work, we presented the results of ab initio calculations of {111} stacking fault energies in γ-TiAl and ternary Ti-rich γ-Ti-Al-X alloys (X being a transition metal element). The energy of the anti-phase boundary of binary γ-TiAl was found to have the highest energy (635 mJ/m^2), followed by the complex stacking fault (370 mJ/m^2) and a superlattice intrinsic stacking fault (173 mJ/m^2). Cross-checking various methodological aspects (exchange-correlation potential, geometry of the supercell, type of relaxation) revealed that the results from various methods are consistent with each other, with the only exception being the scheme for the relaxation of atomic positions (only in the direction perpendicular to the fault plane or full).

Ternary additions, Cr, Hf, Mo, Nb, Ta, V, W and Zr, as well as Ti anti-sites were found to lead to a significant decrease of the SFEs by 20–40%. Such a strong impact is predicted only in the case when the ternary addition is directly in the fault plane, hence corresponding to a composition Ti-50Al-12.5X. The reduction of SFE is much smaller when the alloying element is localized in the plane next to the SF, while it is practically negligible when is it further from the SF plane. In addition to the reduction of the SFEs, our calculations predict stabilization of the APB(111) by Nb, Ti and V, a defect predicted to be unstable in stoichiometric γ-TiAl at 0 K.

Acknowledgments: Financial support from the German BMBF Project O3X3530A and the Austrian Science Fund (FWF) Project Number P29731 is greatly acknowledged.The computational results presented have been achieved in part using the Vienna Scientific Cluster (VSC).The authors are also thankful to Prof. Jörg Neugebauer and his group at the Max-Planck-Institut für Eisenforschung, Düsseldorf, Germany, for their technical support.

Author Contributions: P.D. performed all the ab initio calculations, evaluated results, and did the initial discussion. D.H. supervised the work and prepared this manuscript. H.C. and S.M. initiated the work and contributed to the discussion. All authors participated in writing and shaping this manuscript.

Conflicts of Interest: The authors declare no conflict of interest.

Abbreviations

The following abbreviations are used in this manuscript:

APB	anti-phase boundary
CASTEP	Cambridge Serial Total Energy Package
CPA	coherent potential approximation
CSF	complex stacking fault
DFT	density functional theory
ESF	extrinsic stacking fault
FP-LAPW	full-potential linearized augmented plane-wave (method)
FP-LMTO	full-potential linear muffin-tin orbital (method)
GGA	generalized gradient approximation
GSFE	generalized stacking fault energy
LDA	local density approximation
ISF	intrinsic stacking fault
SF	stacking fault
SFE	stacking fault energy
SISF	superlattice intrinsic stacking fault
TEM	transmission electron microscopy
TM	transition metal
VASP	Vienna Ab initio Simulation Package
xc	exchange and correlation (potential, effects)

References

1. Appel, F.; Paul, J.; Oehring, M. *Gamma Titanium Aluminide Alloys: Science and Technology*; Wiley: Hoboken, NJ, USA, 2011.
2. Clemens, H.; Mayer, S. Design, Processing, Microstructure, Properties, and Applications of Advanced Intermetallic TiAl Alloys. *Adv. Eng. Mater.* **2013**, *15*, 191–215.
3. Mayer, S.; Erdely, P.; Fischer, F.D.; Holec, D.; Kastenhuber, M.; Klein, T.; Clemens, H. Intermetallic β-Solidifying γ-TiAl Based Alloys—From Fundamental Research to Application. *Adv. Eng. Mater.* **2017**, *19*, 1600735.

4.   Dimiduk, D.M. Gamma titanium aluminide alloys—An assessment within the competition of aerospace structural materials. *Mater. Sci. Eng. A* **1999**, *263*, 281–288.
5.   Tadmor, E.B.; Bernstein, N. A first-principles measure for the twinnability of FCC metals. *J. Mech. Phys. Solids* **2004**, *52*, 2507–2519.
6.   Wiezorek, J.M.K.; Humphreys, C.J. On the hierarchy of planar fault energies in TiAl. *Scr. Metall. Mater.* **1995**, *33*, 451–458.
7.   Yoo, M.H.; Fu, C.L. Physical constants, deformation twinning, and microcracking of titanium aluminides. *Metall. Mater. Trans. A* **1998**, *29*, 49–63.
8.   Zhang, W.J.; Appel, F. Effect of Al content and Nb addition on the strength and fault energy of TiAl alloys. *Mater. Sci. Eng. A* **2002**, *329-331*, 649–652.
9.   Yuan, Y.; Liu, H.W.; Zhao, X.N.; Meng, X.K.; Liu, Z.G.; Boll, T.; Al-Kassab, T. Dissociation of super-dislocations and the stacking fault energy in TiAl based alloys with Nb-doping. *Phys. Lett. A* **2006**, *358*, 231–235.
10.   Vitek, V.; Ito, K.; Siegl, R.; Znam, S. Structure of interfaces in the lamellar TiAl: Effects of directional bonding and segregation. *Mater. Sci. Eng. A* **1997**, *239*, 752–760.
11.   Ehmann, J.; Fähnle, M. Generalized stacking-fault energies for TiAl: Mechanical instability of the (111) antiphase boundary. *Philos. Mag. A* **1998**, *77*, 701–714.
12.   Liu, Y.L.; Liu, L.M.; Wang, S.Q.; Ye, H.Q. First-principles study of shear deformation in TiAl and Ti$_3$Al. *Intermetallics* **2007**, *15*, 428–435.
13.   Wen, Y.F.; Sun, J. Generalized planar fault energies and mechanical twinning in gamma TiAl alloys. *Scr. Mater.* **2013**, *68*, 759–762.
14.   Kanani, M.; Hartmaier, A.; Janisch, R. Interface properties in lamellar TiAl microstructures from density functional theory. *Intermetallics* **2014**, *54*, 154–163.
15.   Simmons, J.P.; Rao, S.I.; Dimiduk, D.M. Atomistics simulations of structures and properties of 1/2 dislocations using three different embedded-atom method potentials fit to γ-TiAl. *Philos. Mag. A* **1997**, *75*, 1299–1328.
16.   Mahapatra, R.; Girshick, A.; Pope, D.P.; Vitek, V. Deformation mechanisms of near-stoichiometric single phase TiAl single crystals: A combined experimental and atomistic modeling study. *Scr. Metall. Mater.* **1995**, *33*, 1921–1927.
17.   Kanani, M.; Hartmaier, A.; Janisch, R. Stacking fault based analysis of shear mechanisms at interfaces in lamellar TiAl alloys. *Acta Mater.* **2016**, *106*, 208–218.
18.   Woodward, C.; Maclaren, J.M. Planar fault energies and sessile dislocation configurations in substitutionally disordered Ti-Al with Nb and Cr ternary additions. *Philos. Mag. A* **1996**, *74*, 337–357.
19.   Vitek, V. Theory of the Core Structures of Dislocations in Body-Centered-Cubic Metals. *Cryst. Lattice Defects* **1974**, *5*, 1–34.
20.   Hohenberg, P.; Kohn, W. Inhomogeneous electron gas. *Phys. Rev.* **1964**, *136*, B864–B871.
21.   Kohn, W.; Sham, L.J. Self-consistent equations including exchange and correlation effects. *Phys. Rev.* **1965**, *140*, A1133–A1138.
22.   Kresse, G.; Furthmüller, J. Efficiency of ab-initio total energy calculations for metals and semiconductors using a plane-wave basis set. *Comput. Mater. Sci.* **1996**, *6*, 15–50.
23.   Kresse, G.; Furthmüller, J. Efficient iterative schemes for *ab initio* total-energy calculations using a plane-wave basis set. *Phys. Rev. B Condens. Matter* **1996**, *54*, 11169–11186.
24.   Perdew, J.P.; Wang, Y. Accurate and simple analytic representation of the electron-gas correlation energy. *Phys. Rev. B Condens. Matter* **1992**, *45*, 13244–13249.
25.   Clemens, H.; Mayer, S. Intermetallic titanium aluminides in aerospace applications—Processing, microstructure and properties. *Mater. High Temp.* **2016**, *33*, 560–570.
26.   Jiang, C. First-principles study of site occupancy of dilute 3d, 4d and 5d transition metal solutes in L1$_0$ TiAl. *Acta Mater.* **2008**, *56*, 6224–6231.
27.   Holec, D.; Reddy, R.K.; Klein, T.; Clemens, H. Preferential site occupancy of alloying elements in TiAl-based phases. *J. Appl. Phys.* **2016**, *119*, 205104.
28.   Dumitraschkewitz, P. Planar Faults in γ-TiAl: An Atomistic Study. Master's Thesis, Montanuniversität Leoben, Leoben, Austria, 2015.
29.   Zope, R.R.; Mishin, Y. Interatomic potentials for atomistic simulations of the Ti-Al system. *Phys. Rev. B Condens. Matter* **2003**, *68*, 024102.

30. Farkas, D.; Jones, C. Interatomic potentials for ternary Nb-Ti-Al alloys. *Modell. Simul. Mater. Sci. Eng.* **1999**, *4*, 23.
31. Thornton, P.R.; Mitchell, T.E.; Hirsch, P.B. The dependence of cross-slip on stacking-fault energy in face-centered cubic metals and alloys. *Philos. Mag.* **1962**, *7*, 1349–1369.

applied sciences

MDPI

Article

The Determination of Dendrite Coherency Point Characteristics Using Three New Methods for Aluminum Alloys

Iban Vicario Gómez [1,*], Ester Villanueva Viteri [1], Jessica Montero [2], Mile Djurdjevic [3] and Gerhard Huber [3]

[1] Department of Foundry and Steel Making, Tecnalia Research & Innovation, c/Geldo, Edif. 700, E-48160 Derio, Spain; ester.villanueva@tecnalia.com
[2] Befesa Aluminio, Carretera Lutxana-Asua 13, 48950 Erandio, Spain; jessica.montero@befesa.com
[3] Nemak Linz, Zeppelinstrasse 24, 4030 Linz, Austria; mile.djurdjevic@nemak.com (M.D.); gerhard.huber@nemak.com (G.H.)
* Correspondence: iban.vicario@tecnalia.com; Tel.: +34-943-005-511

Received: 24 May 2018; Accepted: 23 July 2018; Published: 26 July 2018

Featured Application: Increase the accuracy of solidification software for aluminum alloys.

Abstract: The aim of this work is to give an overview of existing methods and to introduce three new methods for the determination of the Dendrite Coherency Point (DCP) for $AlSi_{10}Mg$ alloys, as well as to compare the acquired values of DCP based on a thermal analysis and on the analysis of cooling curves working with only one thermocouple. Additionally, the impact of alloying and contaminant elements on the DCP will be also studied. The first two proposed methods employ the higher order derivatives of the cooling curves. The DCP was determined as the crossing point of the second and third derivative curves plotted versus time (method 1) or that of the temperature (method 2) with the zero line just after the maximum liquidus temperature. The third proposed method is based on the determination of the crossing point of the third solid fraction derivative curve with the zero line, corresponding to a minimum of the second derivative. A Taguchi design for the experiments was developed to study the DCP values in the $AlSi_{10}Mg$ alloy. The DCP temperature values of the test alloys were compared with the DCP temperatures predicted by the previous methods and the influence of the major and minor alloying elements and contaminants over the DCP. The new processes obtained a correlation factor r^2 from 0.954 and 0.979 and a standard deviation from 1.84 to 2.6 °C. The obtained correlation values are higher or similar than those obtained using previous methods with an easier way to define the DCP, allowing for a better automation of the accuracy of DCP determination. The use of derivative curves plotted versus temperature employed in the last two proposed methods, where the test samples did not have an influence over the registration curves, is proposed to have a better accuracy than those of the previously described methods.

Keywords: aluminum alloys; dendrite coherency point; DCP; thermal analysis

1. Introduction

Thin aluminum cast structural parts produced by the Vacuum High Pressure Die Casting (HPDC) process are applied more and more in the automotive industry. Among the many commercial cast aluminum alloys used in HPDC production, the $AlSi_{10}Mg$ alloy has found significant application due to an excellent combination of its high ductility values with a good crush performance of its final cast parts [1].

The solidification of an aluminum alloy begins at the liquidus temperature with the formation of many small crystal nuclei in the molten metal, promoted by melt undercooling. Further cooling leads to a more significant precipitation of the primary dendritic network of α-Al crystals. A dendrite is a tree-like crystal structure that grows in molten metal as the alloy freezes. From a single nucleus, the dendrite grows forward (primary) and laterally (secondary) until the primary dendrite meets another dendrite. The temperature at which this occurs is defined as the dendrite coherency temperature and the solid fraction formed until this temperature is named the dendrite coherency point fraction. The development of the α-aluminum dendritic structure that follows is the growth of the secondary and even tertiary branches with a coarsening of the secondary dendrite arms. Before the molten alloy arrives at the dendrite coherency temperature, the mass feeding of a mixture of the slurry and molten alloy is possible. The impingement of the α-aluminum crystals at the dendrite coherency temperature significantly reduces the flowability of the residual melt and feeding changes from "mass" to inter-dendritic feeding, where the molten metal starts to flow through the solid skeleton of the α-Aluminum dendrites. The solidification of the primary α-aluminum dendrites increases the concentration of the alloying elements in the remaining liquid, promoting the precipitation of AlSi primary eutectic phase, as well other inter-metallics in the hypoeutectic alloys [2]. The major alloying elements have a significant impact on the solidification path of the AlSi alloys, but some minor elements or contaminants can also change the solidification path of those alloys [3]. However, there is a lack of knowledge in the available literature on how different minor alloying elements and contaminants alone or in combination with major alloying elements can impact the DCP temperature in the $AlSi_{10}Mg$ alloy, based on the available methods applied to detect this point.

According to many authors [3–11], the DCP marks the point where casting defects such as shrinkage porosity, hot tearing, and macro-segregation start to appear. A good understanding of the solidification phenomena related to DCP and knowledge of the influence of alloying elements and process parameters on this point are needed for the development of new alloys and, especially, for improving the accuracy of simulation procedures, as well as optimizing HPDC processes.

Thermal analysis (TA) is a quite spread quality control system in aluminum casting plants. The solidification path of molten alloys is plotted in a temperature versus time graph. The obtained curve is called the cooling curve and, together with its derivatives, is employed to characterize the solidification path of different alloys. The existing techniques for determination of the DCP are given below.

Based on the extended literature research [1–24], there are four main processes for the determination of DCP temperature:

1. the mechanical (rheological) method,
2. the two thermocouples method using the minimum temperature difference,
3. the single thermocouple method using the minimum of the second derivative of the cooling curve and/or the common point of the second and third derivative in the zero axis,
4. the three thermocouples method determining the thermal diffusivity during solidification.

The mechanical method monitors the torque required to rotate a disc or a paddle in molten aluminum [8,9] until the shear strength value starts to increase its value at the DCP point at a constant rotation speed.

The two thermocouples technique, or the TA method [10,11], determines the temperature in the center (T_C) of a test crucible and at a nearby inner wall (T_W) using two thermocouples. The DCP temperature is determined by the local minimum on the ΔT versus time curve ($\Delta T = T_W - T_C$) and its projection on the T_C cooling curve. Heat removal from the solid phase is faster than from the liquid phase and occurs at the minimum of the ΔT versus time curve because there is a higher thermal conductivity in the solid dendrites than in the surrounding liquid metal.

Other similar methods based on one thermocouple have been developed to decrease costs and increase productivity in the data analysis by reducing the total amount of processed data.

The first method to define the DCP with one thermocouple located in the center of the TA cup is based on the determination of the first minimum point on the plotted second derivative vs. time graph [12–14], as shown in Figure 1.

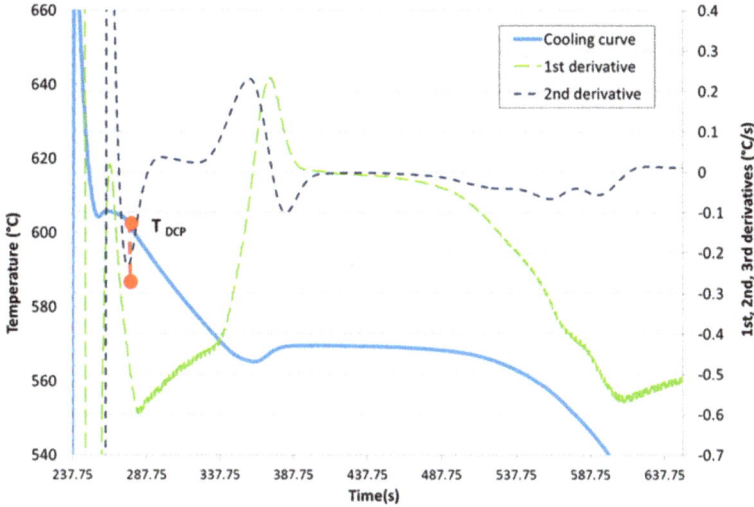

Figure 1. Method 1: the first minimum of the d^2T/dt^2 curve.

The second method with one thermocouple is based on the detection of the first minimum of the first derivative curve plotted vs. time graph [15,16] as shown in Figure 2 with the determination of the maximum liquidus temperature in the first negative crossing of the first derivative curve with the zero line ($T_{liq\,max}$) and the determination of the DCP temperature in the first minimum of the dT/dt curve immediately after the maximum liquidus point.

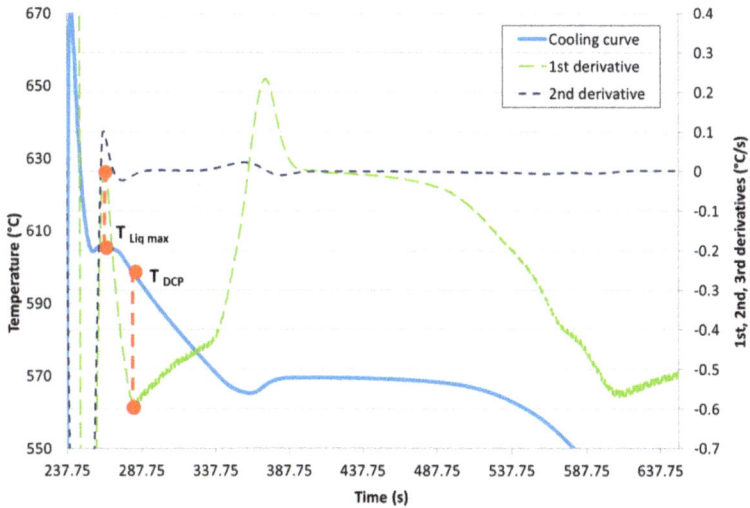

Figure 2. Method 2: the first minimum of the dT/dt curve after $T_{liq\,max}$.

Some works indicate that sometimes the thermal signal is so weak that it is difficult to define the minimum point on the second derivative curve [17].

The third method using only one thermocouple is based on the first derivative curve plotted versus the temperature analysis with the determination of the point at which the first derivative curve starts to change its slope [1], as shown in Figure 3.

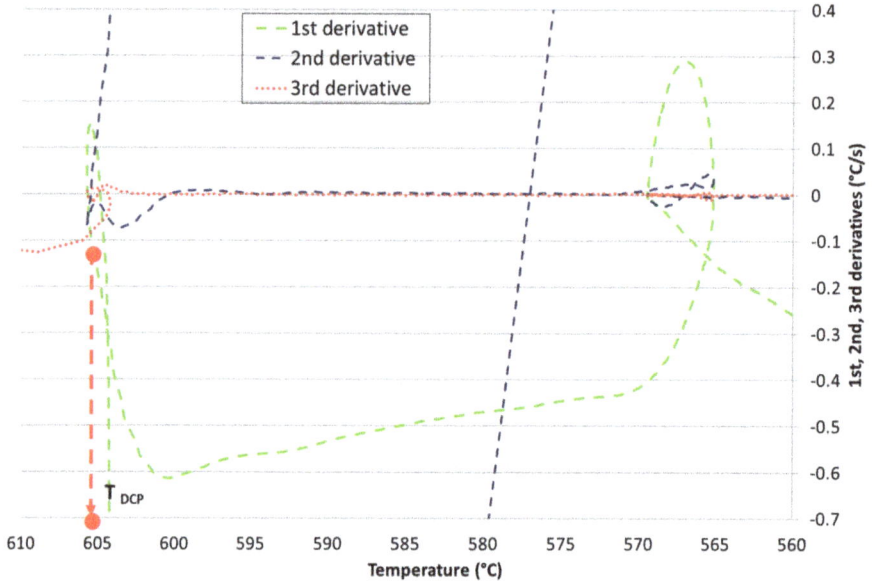

Figure 3. Method 3: dT/dt curve vs. T, with the DCP point in the elbow.

However, it is sometimes difficult to define the exact point of deviation because there are no loops in the first derivative curve, so it is not possible to define the exact position of the elbow point on the dT/dt versus temperature curve.

The solid fraction at the DCP can be determined by using different experimental and/or arithmetic methods [8]. Among them, the Newtonian and Fourier [20–22] methods are mostly applied in the case when the cooling curve data are known.

For the Newtonian analysis, first, the solid fraction at each point or temperature must be calculated, determining the integration or cumulative area between the cooling rate (first derivative (dT/dt) of the cooling curve) and the baseline dT_{BL}/dt (BL). The base line corresponds to a cooling rate curve if there is no phase transformation. Applying this method, it is possible to determine the amount of solid fraction at the dendritic coherence point, identifying the temperature at which this event occurs. This temperature is determined in the elbow of the first derivative of the cooling curve when it starts to be constant. This method is applied as the following Figure 4 exhibits.

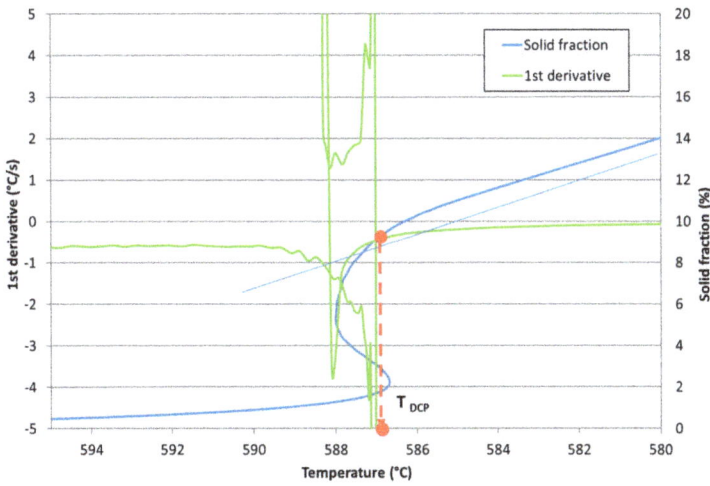

Figure 4. Method 4: dfs/dT curve vs. T, with the DCP point in the elbow.

The determination of the crossing point of the second and third derivative curves plotted versus time after the maximum liquidus point has been proposed as a solution in hypoeutectic ductile iron alloys [23] with only one thermocouple and it is the base for the first proposed method, where the same concept has been employed for hypoeutectic aluminum alloys, as can be observed in Figure 5.

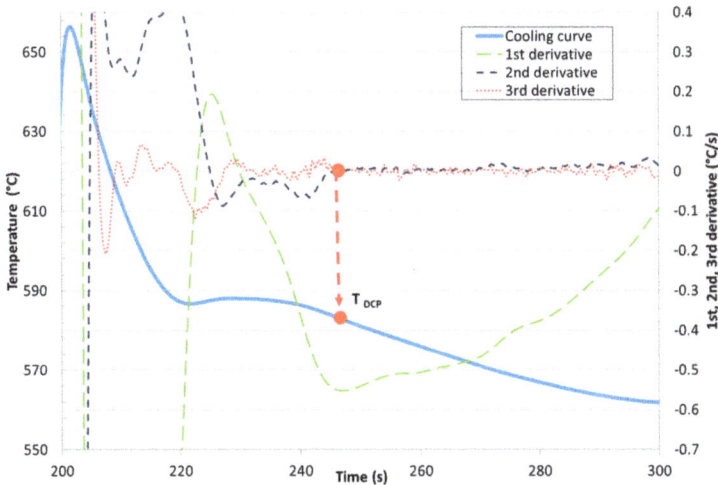

Figure 5. Method 5: the DCP determination at the crossing point of the second and third derivative of dT/dt vs. time.

The three thermocouples method employs thermocouples located at the center of the wall, the middle of the wall, and close to the wall of steel or graphite crucibles, measuring the variation in the thermal diffusivity during the solidification process [24]. We compared this method with the other methods mentioned before concluding that all the previously mentioned methods produced similar results.

This work displays the applicability of all available methods (including the three new methods proposed in this work) for having a pretty accurate trend for the determination of DCP independent on the chemical composition of the investigated alloys. This paper also illustrates that the recording of the solidification temperatures using a single thermocouple can be successfully used to accurately detect the DCP temperature. The proposed methods make the determination of the DCP point easier, especially in the case where there is a lack of information and in the case of the last two proposed methods with no influence of the size of the thermal analysis test samples.

2. Materials and Methods

The approach used in the present work has been based on the identification of the effect of 12 main alloying elements in the solidification parameters through the Taguchi methodology. Two orthogonal matrices were used: an L16 matrix and a modified L8 matrix. The former employs two levels that are related to the maximum and minimum amounts of the alloying element. The modified matrix incorporates intermediate values. To perform a statistical evaluation of results, the Excel software was employed for the determination of the linear regression coefficient (r^2) and the standard deviation (S_{ey}) for the obtained results from the 25 tested alloys. The multiple regression analysis techniques seek to derive a single curve that represents the general trend of the data to make extrapolations beyond the limits of the observed data or interpolations. As much of the equations were obtained with a very limited amount of data (25 alloy compositions), they should be used as trend indicators. It is recommended that at least 100 observations (different alloys) be used to ensure a high degree of accuracy.

The base alloy for the developments has been chosen from the most commonly used alloys for HPDC, and alloying elements were added to the melt to obtain the desired compositions. No grain refining or silicon modification master alloys were added to the melts. The selected alloy is $AlSi_{10}Mg$ according to the standard EN AC-43.400 included in the EN 1706:2010 standard. To determinate the obtained alloy composition, a SPECTROMAXx arc spark OES metal analyzer was used. The obtained compositions are given in Table 1.

Table 1. The compositions of base alloys (mass %).

Ref.	Si	Mg	Fe	Cu	Ni	Cr	Mn	Ti	Zn	Pb	Sn	Sr
[1]	9.00	0.30	0.38	0.03	0.00	0.01	0.34	0.02	0.01	0.00	0.002	0.021
[2]	8.02	0.19	0.29	0.02	0.00	0.01	0.21	0.01	0.00	0.00	0.003	0.003
[3]	8.66	0.14	0.30	0.02	0.00	0.01	0.21	0.20	0.29	0.27	0.039	0.014
[4]	10.01	0.69	0.34	0.02	0.23	0.15	0.67	0.02	0.01	0.00	0.002	0.06
[5]	9.75	0.68	0.34	0.023	0.226	0.145	0.72	0.121	0.347	0.138	0.064	0.055
[6]	8.77	0.15	0.85	0.19	0.21	0.16	0.21	0.12	0.16	0.21	0.073	0.006
[7]	8.43	0.11	0.91	0.19	0.19	0.14	0.18	0.19	0.18	0.19	0.066	0.047
[8]	9.02	0.38	1.05	0.29	0.21	0.07	0.81	0.17	0.06	0.21	0.019	0.048
[9]	9.26	0.56	0.73	0.09	0.001	0.069	0.53	0.024	0.212	0.01	0.002	0.007
[10]	11.65	0.58	0.34	0.199	0.196	0.017	0.302	0.239	0.028	0.073	0.032	0.021
[11]	10.54	0.52	0.34	0.16	0.15	0.02	0.31	0.17	0.23	0.26	0.026	0.053
[12]	11.49	0.40	0.91	0.42	0.00	0.14	0.67	0.23	0.15	0.18	0.04	0.046
[13]	11.60	0.46	0.83	0.18	0.00	0.18	0.74	0.02	0.19	0.23	0.003	0.007
[14]	11.64	0.53	0.96	0.08	0.08	0.16	0.08	0.27	0.13	0.08	0.033	0.01
[15]	11.82	0.52	0.96	0.11	0.11	0.14	0.11	0.11	0.18	0.11	0.046	0.023
[16]	11.41	0.35	0.95	0.27	0.30	0.09	0.69	0.25	0.09	0.25	0.026	0.038
[17]	12.07	0.28	0.83	0.13	0.17	0.03	0.49	0.08	0.02	0.16	0.055	0.033
[18]	10.21	0.278	0.43	0.052	0.001	0.069	0.333	0.021	0.083	0.001	0.002	0.013
[19]	10.37	0.28	0.50	0.11	0.00	0.14	0.44	0.02	0.01	0.00	0.002	0.009
[20]	10.64	0.63	0.41	0.05	0.00	0.07	0.33	0.02	0.10	0.00	0.001	0.013

Table 1. *Cont.*

Ref.	Si	Mg	Fe	Cu	Ni	Cr	Mn	Ti	Zn	Pb	Sn	Sr
[21]	10.31	0.29	0.54	0.09	0.00	0.11	0.35	0.01	0.01	0.00	0.002	0.006
[22]	10.80	0.52	0.48	0.052	0.001	0.064	0.334	0.028	0.095	0.002	0.002	0.014
[23]	10.90	0.43	0.51	0.10	0.00	0.11	0.47	0.01	0.02	0.00	0.005	0.006
[24]	11.71	0.442	0.57	0.073	0.002	0.075	0.438	0.016	0.042	0.002	0.002	0.013
[25]	10.73	0.355	0.6	0.099	0.001	0.087	0.384	0.016	0.102	0.001	0.002	0.009

The procedure to acquire the cooling curve is very simple. Liquid aluminum melt is preheated to approximately 100 °C (720 °C in our case) above its liquidus temperature. To obtain cooling curves by Thermal Analysis (TA), the samples with masses of approximately 300 ± 10 g were poured into calibrate sand cups with a T-type thermocouple placed in the middle of the cup. Temperatures between 630–400 °C were recorded. The data of the TA were collected using a high-speed National Instruments Data Acquisition System linked to a personal computer. Each TA trial was repeated three times. The obtained cooling rate was approximately 3 °C/s.

Development of New Methodologies for the Determination of DCP Temperature

The first proposed method is based on previous work developed for the detection of DCP in hypoeutectic iron alloys [23]. The temperature of the DCP is determined as the crossing point of the second and third derivative curves plotted versus time, with the zero line placed nearly after the maximum liquidus temperature. This point reflects the point where the cooling rate becomes constant. We can observe the determination of the DCP point for the first proposed method in Figure 6.

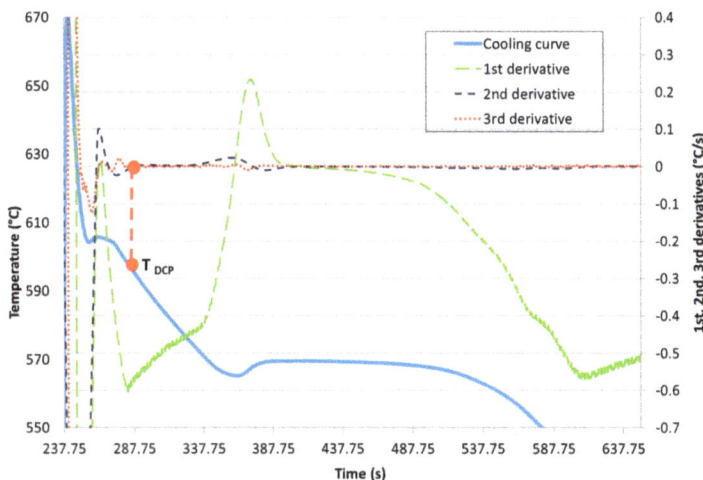

Figure 6. Method 5: the DCP determination in the crossing point of the second and third derivative in the zero axis of the dT/dt curve.

We can observe that the crossing point is closed to the minimum of the first derivative.

The second method is based on the determination of the crossing point of the second and third derivative curves plotted versus temperature with a zero line that corresponds to the DCP. This DCP also reflects the point after which the cooling rate becomes constant. Therefore, in this method, the detection of this point is easier and more accurate compared to the previous methods in which the DCP point was determined at the elbow point of the first derivative curve (dT/dt) with less accuracy. In Figure 7, the determination of the DCP point for the third proposed method is represented.

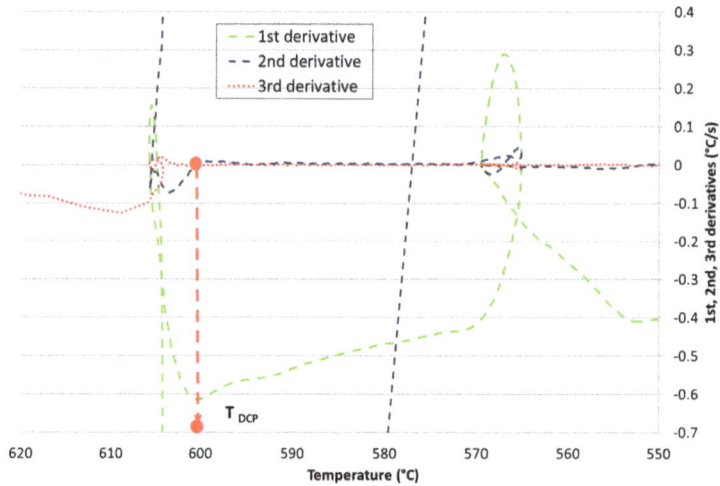

Figure 7. Method 6: the DCP determination at the crossing point of the second and third derivative of dT/dt vs. T curve.

We can observe that the crossing point is close to the minimum of the first derivative.

The third proposed method is based on the determination of the crossing point of the third derivative curve with the zero line of the solid fraction (dFs/dt) plotted versus temperature. This point also corresponds to a minimum in the second derivative. The DCP Temperatures can be determined easily because the size of the thermal analysis test samples does not have as much of an influence over the registration curves as the temperature, which is a thermodynamically extensive property.

We can observe the determination of the DCP point for the second proposed method in Figure 8.

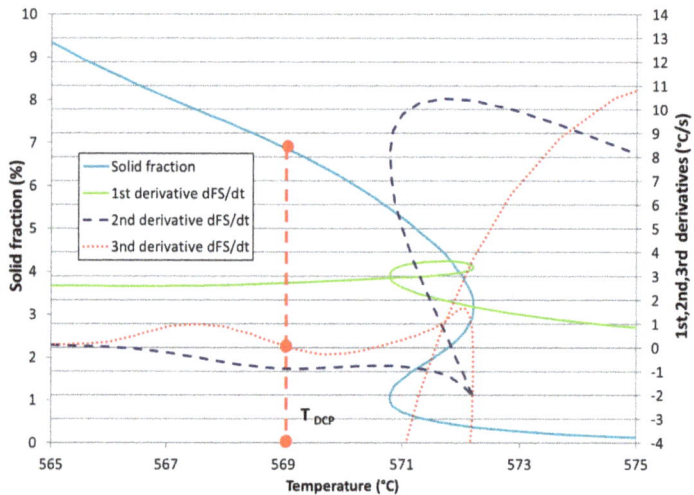

Figure 8. Method 7: the DCP determination at the crossing point of the third derivative of dfs/dt vs. T curve with the zero line.

3. Results

The dendrite coherency temperatures of the studied alloys were determined by applying various methods based on the analysis of the cooling curves and their corresponding derivatives using one thermocouple placed at the center of the test cup. Table 2 summarizes all these temperatures.

Table 2. The DCP temperature values of the studied alloys DCP (°C).

Ref.	Method 1	Method 2	Method 3	Method 4	Method 5	Method 6	Method 7
[1]	590.8	591.6	590.54	591.7	586.82	587.6	590.29
[2]	603.8	599.12	604.46	605.7	599.12	600.47	603.2
[3]	599.82	590.98	599.98	600.0	590.975	590.8	593.31
[4]	585.94	582.47	586.60	587.9	583.03	582.875	585.93
[5]	582.14	575.61	583.19	583.2	575.605	574.945	580.32
[6]	590.25	587.38	594.89	594.9	587.375	587.37	590.2
[7]	598.39	589.62	597.68	598.3	589.615	589.865	593.57
[8]	590.55	586.74	593.53	593.5	586.74	586.72	590.33
[9]	590.8	589.92	592.19	592.1	589.92	588.03	590.56
[10]	576.01	570.32	576.46	576.5	570.32	570.865	575.01
[11]	582.04	573.85	583.05	583.1	573.85	574.46	578.385
[12]	576.17	571.68	576.50	576.5	571.68	571.97	570.425
[13]	572.58	570.01	572.89	573.2	570.005	569.46	572.44
[14]	574.91	568.47	575.32	575.3	568.47	569.085	574.465
[15]	569.96	567.4	570.84	572.0	567.395	567.615	569.405
[16]	575.94	571.3	576.59	576.6	571.3	571.54	571.585
[17]	569.09	567.29	568.79	570.4	567.29	567.82	569.455
[18]	582.8	577.69	583.16	583.8	577.69	578.05	581.815
[19]	583.92	575.97	584.49	585.2	575.965	576.27	583.675
[20]	578.54	575.14	578.23	579.4	575.14	575.77	578.4
[21]	585.08	581.73	586.08	587.2	581.725	582.465	584.585
[22]	577.77	573.61	578.67	579.1	573.61	573.78	576.92
[23]	579.22	576.42	580.98	581.4	576.415	576.7	578.795
[24]	573.8	570.7	575.04	575.2	570.695	570.975	573.2
[25]	579.75	576.17	580.93	581.1	576.17	576.36	579.27

To compare the temperature values obtained for every method and their tendencies, a comparison graph is represented in Figure 9.

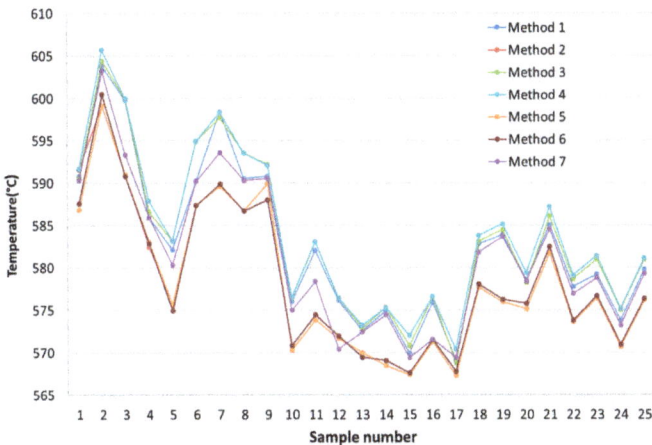

Figure 9. The comparison of the DCP temperatures for every sample with the studied methods.

As it can be observed from Figure 9, all the obtained DCP temperatures could be divided into three groups. The applied methods (methods 2, 5 and 6) detected similar values for DCP. All these values have slightly lower DCP temperatures than those obtained using the other methods. The DCP temperatures detected using methods 1, 3 and 4 are characterized by slightly higher DCP temperatures. The DCP temperatures determined using method 7 are mostly located in the middle, between the two recognized temperature areas. However, it can be observed that all the applied methods are very sensitive to changes in the chemical composition of the investigated alloys.

By using linear regressions calculations with the obtained values of DCP temperatures, Equations (1) to (7) can be written. Some statistical parameters such as the linear regression coefficient (r^2) and the standard deviation (S_{ey}) can also be observed. To define the influence of every alloying element on the studied properties, statistical student t (t) values are employed. The *t*-test is a statistical hypothesis test in which the test statistic follows a Student's t-distribution under the null hypothesis. In our case, the values > 2.66 represent that the selected alloying element has a significant influence over the studied parameters and the "0" values represent that the studied alloying element does not have any influence over the studied parameters (null hypothesis). An intermediate "t" value between 0 and 2.66 shows that the studied parameter has an influence over the DCP temperature, with a higher influence the closer the value is to 2.66, despite not having a statistical influence.

Method 1

$$T_{DCP} \, (^\circ C) = 661.37 - 7.18Si - 6.07Mg - 3.20Fe - 3.33Cu - 6.94Ni + 6.12Cr\text{-}0.59Mn + 23.59Ti - 5.70Zn + 1.69Pb - 53.79Sn + 14.72Sr; \, r^2 = 0.977; \, S_{ey} = 1.99.$$ (1)

Method 2

$$T_{DCP} \, (^\circ C) = 657.2 - 7.60Si + 3.54Mg + 3.14Fe - 3.59Cu\text{-}10.78Ni - 14.20Cr + 2.08Mn + 3.97Ti - 17.71Zn + 5.26Pb + 1.21Sn - 27.38Sr; \, r^2 = 0.976; \, S_{ey} = 2.39.$$ (2)

Method 3

$$T_{DCP} \, (^\circ C) = 665.71 - 7.65Si - 3.16Mg - 3.21Fe + 3.24Cu - 0.62Ni + 7.1Cr - 0.69Mn + 17.84Ti - 5.99Zn + 3.4Pb - 45.53Sn - 32.19Sr; \, r^2 = 0.977; \, S_{ey} = 2.02.$$ (3)

Method 4

$$T_{DCP} \, (^\circ C) = 666.92 - 7.67Si - 2.26Mg - 3.01Fe + 1.76Cu - 3.15Ni + 7.00Cr - 1.77Mn + 13.23Ti - 11.42Zn + 6.33Pb - 34.09Sn - 14.18Sr; \, r^2 = 0.977; \, S_{ey} = 2.05.$$ (4)

Method 5

$$T_{DCP} \, (^\circ C) = 654.69 - 7.38Si + 2.3Mg + 2.30Fe - 0.89Cu - 5.96Ni - 6.15Cr - 9.16Mn + 3.55Ti - 13.41Zn + 3.96Pb - 9.34Sn - 39.17Sr; \, r^2 = 0.\,954; \, S_{ey} = 2.60.$$ (5)

Method 6

$$T_{DCP} \, (^\circ C) = 655.47 - 7.31Si + 2.43Mg + 1.47Fe - 0.86Cu - 7.87Ni - 8.04Cr + 0.16Mn + 4.20Ti - 19.26Zn + 6.69Pb - 7.06Sn - 26.46Sr; \, r^2 = 0.960; \, S_{ey} = 2.46.$$ (6)

Method 7

$$T_{DCP} \, (^\circ C) = 661.64 - 7.55Si + 2.70Mg - 0.25Fe - 7.39Cu - 3.21Ni + 3.46Cr - 0.34Mn + 9.79Ti - 18.85Zn + 5.41Pb - 20.40Sn - 40.57Sr; \, r^2 = 0.979; \, S_{ey} = 1.84.$$ (7)

The student "t" coefficients for temperature DCP obtained by each one of the regressions are shown in the following Table 3.

Table 3. The student "t" coefficients for T_{DCP}.

Meth.	Si	Mg	Fe	Cu	Ni	Cr	Mn	Ti	Zn	Pb	Sn	Sr
1	**13.57**	1.17	1.22	0.39	0.79	0.6	0.18	2.65	0.56	0.16	1.42	0.4
2	**11.95**	0.57	0.99	0.35	1.02	1.17	0.51	0.37	1.45	0.42	0.03	0.61
3	**14.23**	0.6	1.2	0.37	0.07	0.69	0.2	1.97	0.58	0.32	1.18	0.85
4	**14.09**	0.42	1.11	0.2	0.35	0.67	0.51	1.44	1.09	0.6	0.87	0.37
5	**10.65**	0.4	0.67	0.08	0.52	0.69	0.4	0.3	1.01	0.29	0.19	0.8
6	**11.16**	0.38	0.45	0.08	0.72	0.64	0.04	0.38	1.54	0.52	0.15	0.57
7	**15.45**	0.56	0.1	0.94	0.39	0.37	0.11	1.19	2.01	0.57	0.58	1.18

If we obtain a representation of the statistical effect of silicon over the DCP temperature, we can observe that its linear regression coefficient is 0.935, for example, if we employ the calculations of method 7, as shown in Figure 10, it is not as good as the obtained 0.979 value, including the rest of alloy elements.

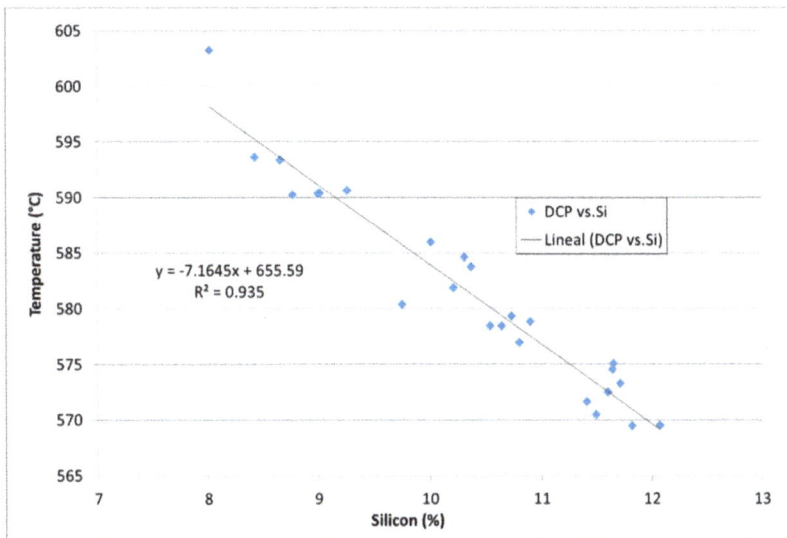

Figure 10. The effect of the Si percentage over the dendrite coherency point temperature with method 7.

4. Discussion

The proposed methods overcame the problems detected to make an accurate determination of the DCP temperature. The determination of the DCP point is very simple and done with a good accuracy in comparison to previous methods, with only one thermocouple, promoting an increased productivity with a low cost in the data analysis.

There is a similar tendency in the DCP values in all the methods in relation to the variation of the alloy composition. Methods 1, 3, and 4 show a tendency to have higher DCP temperatures values because the acceleration of the cooling rate is the basis of defining the exact point at which the DCP starts in these methods, where a limited number of dendrites touch one to another, but promote the increase of the cooling rate in the sample. The rest of the methods are based on the determination of the exact moment when the cooling speed is constant, so all the dendrites touch one to another.

From the studied methods, we estimate that method 6 and method 7 could be the ones with the better trend accuracy due to the fact that the use of the derivative curves plotted versus temperature is not as influenced by the size of the thermal analysis test samples on the registration curves as the temperature, which is a thermodynamically extensive property. Additionally, very similar values were obtained from both methods. In the case of method 3, where the dT/dt curve is plotted vs. T with the DCP point in the elbow, and method 4, in which the dfs/dT curve vs. T exists with the DCP point in the elbow, there is no clear indication of which one is the exact point and it is also very complicated to obtain a curve showing a perfect elbow.

The only element that has a direct effect in the studied alloys with a significant statistical influence over the TDCP is Si, with a lower impact than the other alloying elements. The increase in the percentage of Si promotes a decrease in TDCP. This is an expected behavior because it is well known that an increase in the Si% decreases the solidification interval of hypoeutectic aluminum alloys and their related characteristic solidification temperatures until the minimum solidification temperature interval is reached by eutectic composition [24]. The obtained formulae should be taken as trend indicators [25] and, taking this into account, Ni and Zn show a tendency to decrease the TDCP. It is known that an increase in the Zn% decreases the characteristic solidification temperatures of hypoeutectic aluminum alloys because the Zn enters into the solid dissolution in the alloy matrix and not into the grain boundary, avoiding the enrichment of Zn into the remaining inter-dendritic liquid metal. The decrease of the TDCP is not as expected in the case of Ni and it could be related to the formation of Al_3Ni intermetallic compounds that are precipitated in the beginning of the solidification process of the alloy, at temperatures well above the TDCP and as described in Reference [26] because Ni provides significant changes in the sequence of post-eutectic reactions, promoting a substantial reduction in the alloy's freezing range. In both cases, the obtained results confirm the results obtained for the development of the Si equivalent method for obtaining the solidification temperatures, where Ni and Zn have a positive value, which means that they have an influence on decreasing the solidification temperatures [27,28].

Ti is usually employed in the aluminum industry because it promotes the grain refinement of the aluminum alloys. If the grain is smaller, there are more dendrites in the solidification process, so their tips could touch one to another quicker, increasing the T_{DCP} value, but without statistical relevance. The obtained results could also be correlated with the previous studies so that they show that an alloy refined with Ti has higher solidification temperatures than the unrefined alloys [15,22]. Pb is usually precipitated in the grain boundary as isolated points and has a very restricted solid dissolution in the aluminum matrix. Because of this, Pb could tend to increase the T_{DCP} value, but also without a statistical relevance. This result is also in concordance with a previous study [27,29], where elements such as grain refiners (Ti and B) and silicon modifiers (Sr and Sb) or elements with a low melting point (Bi and Pb) have similar effects on the Si Equivalent value.

The rest of the alloying elements also have a slight influence on the solidification temperature interval, but it is not very important and there is a complex interaction between them, obtaining a better adjustment of the results by adding all the alloying elements. The difference in the increase or decrease of the rest of the alloying elements can be related to the presence of intermetallic or eutectic compounds. If they precipitate before the DCP, they would decrease the T_{DCP}. Many of the alloying element could precipitate in different inter-metallics and eutectics (For example the Fe as Al_5FeSi, $Al_8FeMg_3Si_6$, and others).

By comparing the studied methods, the linear regression coefficient (r^2) and the standard deviation (S_{ey}) show that in all the cases, a good correlation between the developed formulae and the obtained results in r^2 values > 0.95 and Sey from 1.84 to 2.6 °C.

More investigations with torque measurements should be done in order to define which one of the proposed methods is more exact in real DCP point determination and in the correlation between the different quantities of inter-metallics, types, and concentrations to have an estimation of the influence of the different inter-metallics over the T_{DCP}.

5. Conclusions

A Taguchi based methodology has been employed to calculate the DCP and its temperature. The obtained results presented in this paper show the importance of the composition of the alloy over the DCP temperatures and the differences over the different calculation methods. The results show that the obtained equations allow us to define, with good accuracy, the DCP point of any alloy of the $AlSi_{10}Mg$ family, with a good statistical correlation between the obtained values from the different methods, especially with the newly developed methods.

Silicon is the element with the main influence over the DCP point value, but the rest of the alloying elements, despite not having a statistical signification, have an influence over the final DCP temperature.

The determination of the DCP point employing the point where the second and the third derivative crosses after the maximum liquidus temperature point allows us to obtain, in an easier way, the exact DCP point. Additionally, in the case of employing the determination with the dfs/dt vs. T curve, from the developed new methods, the two based on plotting derivatives versus temperature are supposed to obtain the DCP with a higher accuracy than those obtained by previous methods. These techniques allow for a better automatization of the DCP point determination to be used with TA equipment and simulation software with a reduced cost using only one thermos-couple.

Further studies could correlate the obtained values with the Thermocalc software calculated values, not only for the DCP but also for the Solidification fraction with more alloy test to increase the accuracy of the results will be developed. Additionally, the improvement of solidification simulation software and the calculation of DCP with different alloy compositions will be developed by the mechanical (rheological) method and by the two thermos-couple methods.

Author Contributions: I.V.G., E.V.V. and J.M. conceived and designed the experiments; I.V.G. and E.V.V. performed the experiments; E.V.V. prepared the data and I.V.G., E.V.V., J.M., M.D. and G.H. analyzed the data; I.V.G. wrote the paper. Authorship must be limited to those who have contributed substantially to the work reported.

Acknowledgments: This work has been partially funded by the Basque Government through the ETORGAI programme ZE-2016/00018 and from the European Union's Seventh Programme for research, technological development and demonstration under grant agreement No. 296024.

Conflicts of Interest: The authors declare no conflict of interest.

References

1. Voncina, M.; Mrvar, P.; Medved, J. Thermodynamic analysis of $AlSi_{10}Mg$ alloy. *RMZ M&G* **2006**, *52*, 621–633.
2. Djurdjevic, M.B.; Huber, G. Determination of rigidity point/temperature using thermal analysis method and mechanical technique. *J. Alloys Compd.* **2014**, *590*, 500–506. [CrossRef]
3. Djurdjevic, M.B.; Stockwell, T.; Sokolowski, J. The effect of Strontium on the. microstructure of the Aluminium-silicon and Aluminium-copper eutectics in the 319 Aluminium alloy. *Int. J. Cast Met. Res.* **1999**, *12*, 67–73. [CrossRef]
4. Arnberg, L.; Chai, G.; Bäckerud, L. Determination of dendritic coherency in solidification melts by rheological measurements. *Mater. Sci. Eng. A* **1993**, *173*, 101–103. [CrossRef]
5. Chai, G.; Backerud, L.; Avnberg, L. Study of dendrite coherency in Al-Si alloys during equiaxed dendritic solidification. *Z. Metalk.* **1995**, *86*, 54–59.
6. Chai, G.; Bäckerud, L.; Rolland, T.; Arnberg, L. Dendrite coherency during equiaxed solidification in binary aluminum alloys. *Met. Mater. Trans. A* **1995**, *26*, 965–970. [CrossRef]
7. Veldman, N.; Dahle, A.; St. John, D. Determination of dendrite coherency point. In Proceedings of the Die Casting & Tooling Technology Conference, Melbourne, Australia, 22–25 June 1997; pp. 22–25.
8. Bäckerud, L.; Chai, G.; Tamminen, J. *Solidification Characteristics of Aluminum Alloys. Vol. 2. Foundry Alloys*; American Foundrymen's Society. Inc.: Oslo, Norway, 1990; p. 266.
9. Claxton, R. Aluminum alloy coherency. *J. Miner. Met. Mater. Soc.* **1975**, *27*, 14–16. [CrossRef]

10. Bäckerud, L.; Chalmers, B. Some aspects of dendritic growth in binary alloys: study of the Aluminum–Copper system. *Trans. Met. Soc. AIME* **1969**, *245*, 309–318.

11. Tamminen, J. Thermal Analysis for Investigation of Solidification Mechanisms in Metals and Alloys. Ph.D. Thesis, Stockholm University, Stockholm, Sweden, January 1988.

12. Jiang, H.; Kierkus, W.T.; Sokolowski, J.H. Dendrite coherency point determination using thermal analysis and rheological measurements. In Proceedings of the International Conference on Thermophysical Properties of Materials (TPPM), Singapore, 17–19 November 1999.

13. Djurdjevic, M.B.; Kierkus, W.T.; Sokolowski, J.H. Detection of the dendrite coherency point of Al 3XX series of alloys using a single sensor thermal analysis technique. In Proceedings of the 40th Annual Conference of Metallurgists of CIM, Toronto, ON, Canada; 2001.

14. Djurdjevic, M.B.; Kierkus, W.T.; Liliac, R.E.; Sokolowski, J.H. Extended analysis of cooling curves. In Proceedings of the 41st Annual Conference of Metallurgists of CIM, Montreal, QC, Canada, 11–14 August 2002.

15. Pelayo, G.; Sokolowski, J.H.; Lashkari, R.A. Case based reasoning aluminium thermal analysis platform for the prediction of W319 Al cast component characteristics. *J. Achiev. Mater. Manuf. Eng.* **2009**, *36*, 7–17.

16. Pavlovic-Krstic, J. Impact of Casting Parameters and Chemical Composition on the Solidification Behaviour of Al-Si-Cu Hypoeutectic Alloy. Ph.D. Thesis, Universität Magdeburg, Magdeburg, Germany, 2010.

17. Djurdjevic, M.B.; Sokolowski, J.H.; Odanovic, Z. Determination of dendrite coherency point characteristics using first dericative curve versus temperature. *J. Therm. Anal. Calorim.* **2012**, *109*, 875–882. [CrossRef]

18. Jiang, H.; Kierkus, W.T.; Sokolowski, J.H. Determining dendrite coherency point characteristics of Al alloys using single-thermocouple technique. *Trans. Am. Foundrymen's Soc.* **1999**, *68*, 169–172.

19. Djurdjevic, M.B.; Vicario, I. Description of hypoeutectic Al-Si-Cu Alloys based on their known chemical compositions. *Rev. Metal.* **2013**, *49*, 161–171.

20. Hosseini, V.A.; Shabestari, S.G. Study on the eutectic and post-eutectic reactions in LM13 aluminum alloy using cooling curve thermal analysis technique. *J. Therm. Anal. Calorim.* **2016**, *124*, 611–617. [CrossRef]

21. Marchwica, P.; Sokolowski, J.H.; Kierkus, W.T. Fraction solid evolution characteristics of AlSiCu alloys—Dynamic Baseline Approach. *J. Achiev. Mater. Manuf. Eng.* **2011**, *47*, 115–136.

22. Anjosa, V.; Deike, R.; Silva Ribeiro, C. The use of thermal analysis to predict the dendritic coherency point on nodular cast iron melts. *Ciênc. Tecnol. Mater.* **2017**, *29*, 27–33. [CrossRef]

23. Zamarripa, R.C.; Ramos-Salas, J.A.; Talamantes-Silva, J.; Valtierra, S.; Colas, R. Determination of the dendrite coherency point during solidification by means of thermal diffusivity analysis. *Met. Mater. Trans. A* **2007**, *38*, 1875–1879. [CrossRef]

24. Rana, R.S; Purohit, S.; Das, S. Reviews on the influences of alloying elements on the microstructure and mechanical properties of aluminum alloy composites. *IJSRP* **2012**, *2*, 1–7.

25. Makhlouf, M.; Apelian, D.; Wang, L. Microstructures and Properties of Aluminum die Casting Alloys. 1998. Available online: http://www.osti.gov/biblio/751030 (accessed on 23 July 2018).

26. Rakhmonov, J.; Timelli, G.; Bonollo, F. Characterization of the solidification path and microstructure of secondary Al-7Si-3Cu-0.3Mg alloy with Zr, V and Ni additions. *Mater. Charact.* **2017**, *128*, 100–108. [CrossRef]

27. Hernandez, F.C.R.; Djurdjevic, M.B.; Kierkus, W.T.; Sokolowski, J.H. Calculation of the liquidus temperature for hypo and hypereutectic aluminum silicon alloys. *Mater. Sci. Eng. A* **2005**, *396*, 271–276. [CrossRef]

28. Djurdjevic, M.B.; Francis, R.; Sokolowski, J.H.; Emadi, D.; Sahoo, M. Comparison of different analytical methods for the calculation of latent heat of solidification of 3XX aluminum alloys. *Mater. Sci. Eng. A* **2004**, *386*, 277–283. [CrossRef]

29. Djurdjevic, M.B.; Tekfak, J.; Odanovic, Z. Applications de l'analysis thermique dans les fonderies d'aluminium. *Fonderie* **2012**, *26*, 31–37.

applied
sciences

MDPI

Article

Effects of Shot-Peening and Stress Ratio on the Fatigue Crack Propagation of AL 7475-T7351 Specimens

Natália Ferreira [1], Pedro V. Antunes [1,*], José A. M. Ferreira [1], José D. M. Costa [1] and Carlos Capela [1,2]

[1]   Department of Mechanical Engineering, Centre for Mechanical Engineering, Materials and
      Processes (CEMMPRE), University of Coimbra, Rua Luís Reis Santos, Coimbra 3030-788, Portugal;
      talitasmferreira@gmail.com (N.F.); martins.ferreira@dem.uc.pt (J.A.M.F.);
      jose.domingos@dem.uc.pt (J.D.M.C.); carlos.capela@ipleiria.pt (C.C.)
[2]   Department of Mechanical Engineering, Instituto Politécnico de Leiria, ESTG, Morro do Lena—Alto Vieiro,
      Leiria 2400-901, Portugal
*   Correspondence: pedro.antunes@dem.uc.pt; Tel.: +351-239-790-700

Received: 23 January 2018; Accepted: 28 February 2018; Published: 5 March 2018

Featured Application: 7475-T7351 aluminum alloy are widely used for structural components in aerospace applications.

Abstract: Shot peening is an attractive technique for fatigue enhanced performance of metallic components, because it increases fatigue crack initiation life prevention and retards early crack growth. Engineering design based on fatigue crack propagation predictions applying the principles of fracture mechanics is commonly used in aluminum structures for aerospace engineering. The main purpose of present work was to analyze the effect of shot peening on the fatigue crack propagation of the 7475 aluminum alloy, under both constant amplitude loading and periodical overload blocks. The tests were performed on 4 and 8 mm thickness specimens with stress ratios of 0.05 and 0.4. The analysis of the shot-peened surface showed a small increase of the micro-hardness values due to the plastic deformations imposed by shot peening. The surface peening beneficial effect on fatigue crack growth is very limited; its main effect is more noticeable near the threshold. The specimen's thickness only has marginal influence on the crack propagation, in opposite to the stress ratio. Periodic overload blocks of 300 cycles promotes a reduction of the fatigue crack growth rate for both intervals of 7500 and 15,000 cycles.

Keywords: aeronautical aluminum alloys; fatigue crack propagation; overloads; shot peening; Paris law

1. Introduction

High-strength aluminum alloys are widely used in aerospace applications due to the high strength-to-weight ratio, good corrosion resistance and high toughness combined with good formability and weldability. High-strength aluminum alloys are broadly employed in aerospace applications owing to the high strength-to-weight ratio, excellent corrosion resistance and great toughness associated with good weldability capabilities and formability. On the other hand, one of the main issues for the contemporary aircraft industry is to ensure simultaneously reliability, high durability, minimum weight and economic efficiency of transport aircraft. To obtain such crafted aircraft characteristics, it is necessary to design structures which ensure high damage tolerance. The approach to engineering design based on the assumption that flaws can exist in any structure and cracks propagate in service, is commonly used in aerospace engineering. Therefore, the prediction of crack growth rates based on the application of fracture mechanics theory is an important aspect of a structural damage tolerant assessment.

Many metal components, such as turbines blades, used in aerospace and power industries are subjected to dynamic mechanical loading, leading to the initiation of fatigue cracks. One way to reduce the risk of fatigue crack initiation is to introduce compressive stresses in the region of higher stresses concentration, for example by shot peening. At the industrial level, this is a well-established surface treatment technology, despite generating a meaningful rougher surface and therefore surface defects [1]. Considering that most fatigue cracks initiate at the surface, the conditioning of the surface to resist crack initiation and earlier crack growth is a convenient method to enhance fatigue performance. The indentation of each impact, in shot peening process, produces local plastic deformation given rise to a field of surface compressive stresses. Studies by many researchers have shown a positive shot peening effect [2–4], resulting from the introduction of residual compressive stresses in the subsurface layers of material. Depending on the peened material, there is an Almen intensity for which the optimum fatigue strength is achieved, corresponding to a certain balance between residual compressive stress field and surface roughness damage.

Paris's law, which relates fatigue crack growth rate (da/dN) and stress intensity range (ΔK) is the prime approach adopted for characterizing fatigue crack propagation in engineering structures. Fatigue crack growth of the aluminum alloys in the Paris's law regime is affected by microstructure [5–7] and by the crack closure induced by plasticity, oxidation and surface roughness, especially near threshold regime. Paris law characterizes the rate of crack advance per cycle: $da/dN = C\Delta K^m$. The rate of crack advance per cycle is related to the stress intensity factor range ΔK. C and m are constants that depend on the material, environment and stress ratio. Crack closure is considered a very good approach to explain the influence of mean stress on the fatigue crack growth rate [6,8]. Bergner and Zouhar [6] showed that crack growth rates of various aluminum alloys varied by a factor of about 20 for some ΔK values, suggesting that the main factor to explain that discrepancies was the crack closure effect and the environment. Fatigue cracks tends to grow into a material region which has experienced large plastic strains due to its location in the crack tip plastic zone. Typically, this material is deformed beyond its elastic domain in the direction normal to the crack flanks. The trace of the plastic deformation produced is left in the crack's path. It acts in the same way to an additional wedge stick between flanks, thus pre-straining them and partly protecting the crack tip from the action of posterior loads. This phenomenon is called plasticity induced crack closure and tends to decrease the effective stress intensity range thereby resulting in slower crack propagation rates [8].

In compact tension (CT) specimens, the crack progresses more rapidly in center than at the surface conducting to a crack tunneling effect, due to the prevailing tri-axial state of stress at the center, promoting plain strain in contrast with plain stress at surface. Striation spacing between beach marks on fatigue crack surfaces is also affected by shot and laser peening effect. Zhou et al. [9] have observed a decrease in striation spacing with increase in the number of laser peening impacts for Ti6Al4V specimens'. For the same alloy, Pant et al. [10] studied the effect of shot peening and laser peening on the fatigue crack propagation and compared with the untreated one with respect to the striation spacing; this was done using $R = 0.1$ and $R = 0.7$. Both peening surfaces presented a reduction on the striation spacing when compared to the untreated specimens.

Overloads can lead to significant interaction effects on crack propagation, as has been reported in many studies [11–22]. Crack growth retardation can be explained by many mechanisms, including models based on crack closure, residual stresses, crack tip blunting, strain hardening, reversed yielding and crack branching. The residual plastic deformation effect leads to compressive stresses in the wake of the crack and raises the crack opening load on subsequent crack growth (crack closure), becoming the most important phenomena for what concerns the explanation for the variation of characteristic features of post-overload transients [19–23]. Donald and Paris [23] observed for 6061-T6 and 2024-T3 aluminum alloys that closure measurements produced good data correlation between distinct stress ratio crack growths obtained in tests with increasingly K conditions. However, in the near-threshold regime with crack growth data obtained by the K-decreasing method, measured opening loads were excessive. This discrepancy was justified by Paris et al. [24], who suggested the "partial closure model". Borrego et al. [25] concluded that crack closure explained the bias of stress ratio on the fatigue crack growth rate for the 6082-T6 aluminum alloy and the influence of several load parameters for overloads interactions if the partial crack closure model is included in the analyses.

The present work analyzes the effect of shot peening, specimen thickness and stress ratio on the fatigue crack propagation of 7475 aluminum alloy with T7351 heat treatment. T7351 provides an aged material abler to resist to stress-corrosion, the heat-treatment produces stress-relieved by control stretching and after artificially overaged to achieve the best stress corrosion resistance. A more extensive analysis of the crack growth following periodical tensile overloads blocks is also evaluated.

2. Materials and Experimental Procedures

2.1. Materials and Samples

This research was conducted using the 7475 aluminum alloy with a T7351 heat treatment. These alloys are widely used in aeronautical applications where the combination of high strength, fracture toughness, good fatigue crack propagation and corrosion resistance are required. The chemical composition is shown in Table 1. The material bars from which the specimens were produced had following the dimensions in mm: 4000 × 1000 × 250. According to the material manufacturer, the ultimate tensile stress and yield stress are σ_{UTS} = 490 MPa and σ_{YS} = 414 MPa, respectively.

Table 1. Chemical composition of the 7475-T7351 aluminum alloy (% Weight).

Si	Fe	Cu	Mn	Mg	Cr	Zn	Ti	Others	Al
0.1	0.12	1.2–1.9	0.06	1.9–2.6	0.18–0.25	5.2–6.2	0.06	0.15	Remaining

To study the surface shot peening effect on crack propagation, two CT specimen batches were prepared: one with shot peening (SP) and another without shot peening but with the lateral surfaces mechanically polished (MP).

MP specimens were polished to ensure a good visualization of crack propagation. Manual grinding was done with a LaboPol-5—Struers A/S, DK-2750, machine passing progressively the grinding papers 240, 320, 600, 1000 and 2500. After specimen grinding, diamond paste of 3 μm and 1 μm were used to give the specimens a mirrored surface aspect.

Shot peening was done at OGMA Indústria Aeronáutica de Portugal S.A. Company (Alverca do Ribatejo, Portugal) with a large experience in producing components and aeronautics repair. Both sides of the specimen were subjected to a manual shot peening process, using a SURFATEC machine, asshown in Figure 1a, and an Almen strip type A, according to SAE J443 standard [26]. Coverage assessment was done by the surface visual inspection, using a 10× magnifying lens. One hundred-percent coverage was achieved when this analysis showed a completely attained surface by particles. The beads used in current study were of the type S170 with 0.43 mm diameter and Almen type A with intensity 0.20 A (mm), according SAE AMS2430 standard [27] for aluminum alloys. Figure 1b shows one sample after the shot-peening process.

The studies of fatigue crack propagation were performed using the standard Compact Tension (CT) specimen with the geometry shown in Figure 1c, according with ASTM E647 standard [28]. For each batch of specimens, two different thickness (B) were manufactured: 4 mm and 8 mm. The specimens were machined in the longitudinal transverse (LT) direction from laminated plates. The loading specimen's direction coincides with bars lamination direction (Figure 1b,c). For each test condition, three specimens were used.

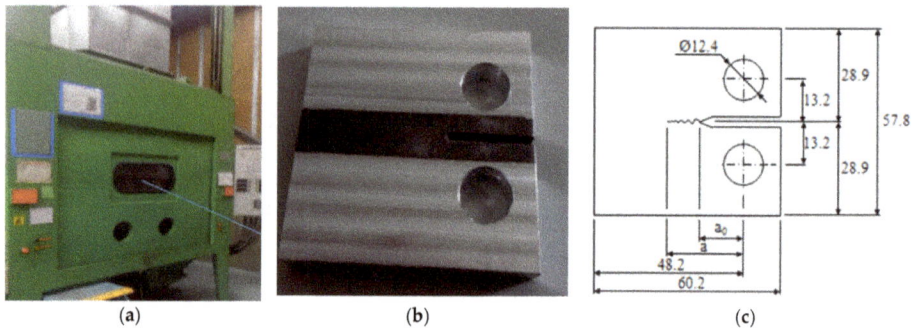

Figure 1. (a) Shot peening machine; (b) shot peened specimen; and (c) dimensions of Compact Tension (CT) specimens in mm.

The surface roughness was evaluated according to DIN EN ISO 4288 standard [29] using a Surftest SJ-500 Mitutoyo, surface roughness measuring system. The evaluated parameters for each superficial treatment were: roughness average R_a, root mean square (RMS) roughness R_q and mean roughness depth R_z. Table 2 summarizes the roughness parameters showing an increasing of more than 300% in the three roughness parameters for the peened surfaces.

Table 2. Surface roughness parameters for Mechanically Polished (MP) and Shot Penned (SP) specimens.

Specimen	Parameter	Mean Value ± Standard Deviation (μm)
MP	Ra	1.22 ± 0.02
	Rq	1.50 ± 0.02
	Rz	7.74 ± 0.13
SP	Ra	3.70 ± 0.17
	Rq	4.60 ± 0.21
	Rz	23.50 ± 2.00

To analyze the material microstructure, some samples of the specimens were selected to observe their cross section. The specimen's surface was gradually polished with several silicon carbide papers. The papers' granulometry ranged from high to low. Afterwards, 1 μm diameter diamond particles were used until specimen's surface became mirror-like. Surfaces were then etched with Keller reagent (2.5% HNO_3, 1.5% HCl, 1% HF, and 95% H_2O (volume) (Coventry, UK)) and taken micrographs using an optical microscope Leica DM 4000 M LED (Wetzlar, Alemanha). Figure 2 shows typical micrographs indicating that base material microstructure (Figure 2a) with elongated grains in the rolling direction. The plane selected to take micrograph was normal to the loading direction to demonstrate the shot-peening effect. Around the shot peened surface (Figure 2b), an increasing of grain deformation and roughness was observed.

Surface Vickers hardness tests were performed according to ASTM C1327-15 [30] using a Struers Duramin micro-hardness tester with 0.5 N load for 15 s. Micro-hardness measurements were done in the cross-section of the sample at 0.3 mm from the surface, and spaced out 0.5 mm, for both specimen types, MP and SP. The average values obtained from twenty measurements were: $HV_{0.05} = 157$ for MP surfaces and $HV_{0.05} = 167$ for SP surfaces. Therefore, shot peening surface hardness increased is more than 6%.

Residual stresses were measured, in-depth and on the longitudinal surface. Residual stresses analysis was performed by X-ray diffraction using a Proto iXRD equipment. Lattice deformations of the {222} diffraction planes were measured using Cr-Kα X-ray radiation, with 22° ψ angles, in the range ±42°, an acquisition time of 30 s by peak and ±2° oscillations in ψ. For the analyzed material, and considering the radiation used, the average penetration depth of the X-rays was about 11 μm. Measurements were made for all studied treatments at four points along the surface of one specimen: one point for each longitudinal surface and one point for each notch surface in the central position. The analysis of the in-depth evolution of the residual stresses in the longitudinal surface was performed by X-Ray diffraction after successive layer removal by electro polishing.

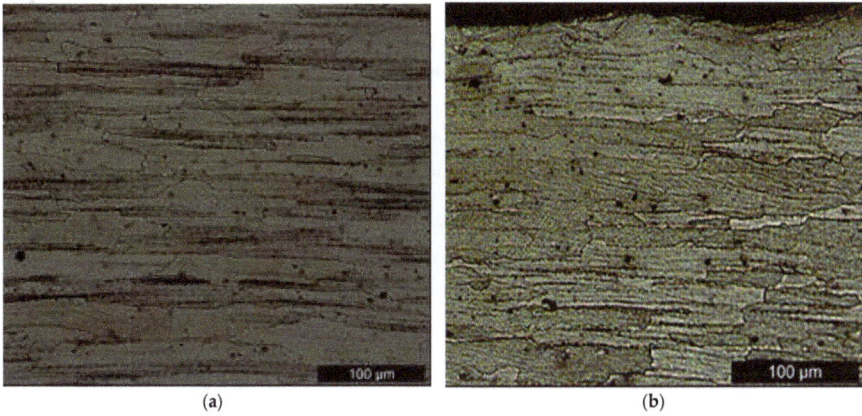
(a)   (b)

Figure 2. Microstructure micrographs: (**a**) base material; and (**b**) shot peening (SP) sample.

2.2. Fatigue Tests

Fatigue crack propagation tests were carried out, in agreement with ASTM E647 standard [28], using 4 and 8 mm thick compact specimens (CT). The tests were performed under load control at room temperature using a 100 kN capacity servo-hydraulic Instron 1341, with a frequency within the range 15–20 Hz and stress ratios of $R = 0.05$ and 0.4. The specimen's geometry and dimensions are shown in Figure 1c. For both specimens' batches two types of tests were conducted: constant amplitude loading tests with the stress ratios $R = 0.05$ and 0.4 and variable amplitude loading tests in which periodic overload blocks of 300 cycles are applied with intervals of N_{int} cycles, as shown schematically in Figure 3. The main purpose of these tests is to obtain the a-N and da/dN curves as a function of the stress intensity factor range ΔK to analyze the effects of the shot peening, specimen thickness and stress ratio.

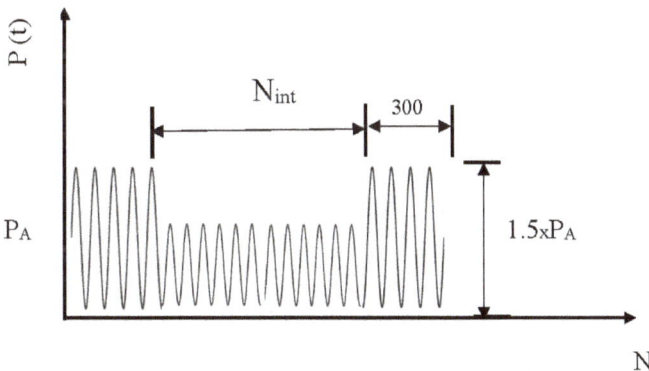

Figure 3. Scheme of variable amplitude loadings with periodic overloading blocks.

For the surface polished specimens, the surface crack length was measured using a travelling microscope (45×) with 10 µm accuracy. Crack growth rates under constant amplitude loading were determined by the incremental polynomial method using five consecutive points [28]. For the surface peened specimens, the crack length was obtained by using experimental calibration curves based on the compliance variation, previously obtained with the polished specimen's tests, considering the compliance (*C*) definition and ratio of displacement to load increment (Equation (1)).

$$C = \frac{(u_{máx} - u_{mín})}{(P_{máx} - P_{mín})} \qquad (1)$$

where *u* and *P* are the axial grip displacement and the load, respectively, which were monitored during the test. From the non-peened specimen tests with constant amplitude loading, it was monitored a set of data for *C* calculation and the correspondent values of the crack length. The collected data are plotted in Figure 4, in terms of the crack length (*a* in mm) versus the compliance, and fitted by Equations (2) and (3) for specimens with 8 mm and 4 mm thickness, respectively, both with a 0.99 correlation factor. Equations (2) and (3) were afterwards used for the evaluation of the crack length in the tests with peened specimens and in the periodical overloading block tests.

$$a = 39,515 \times C^5 - 53,199 \times C^4 + 27,962 \times C^3 - 7278.2 \times C^2 + 980.36 \times C - 23.559 \qquad (2)$$

$$a = 2905.2 \times C^5 - 6399 \times C^4 + 5393.9 \times C^2 + 2275.3 \times C^2 + 502.82 \times C - 16.62 \qquad (3)$$

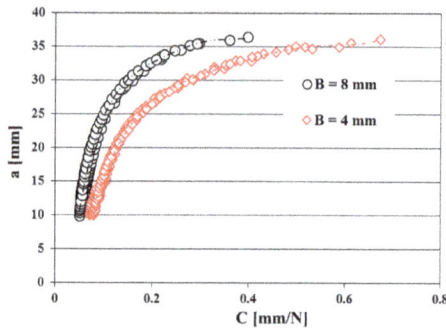

Figure 4. Calibration curves based on the compliance, *C*, for 4 and 8 mm thick specimens.

3. Results and Discussion

Figures 5 and 6 highlight the effects of the specimen's thickness and surface peening on the crack propagation curves, respectively. Figure 5a–d shows the influence of the thickness on the *da/dN*-Δ*K* curves. It is well known [31] that the thickness influence on the fatigue crack propagation is related both to the microstructure and stress state. In the present study, specimens were machined from the same thickness bars, so there is no microstructure change between 4 and 8 mm thickness specimens. Therefore, the effect of thickness is only caused by changes in stress distribution along cross section and consequent variation on crack closure level [31]. The analysis of Figure 5 shows a reduced thickness effect on *da/dN* for both surface treatments (MP and SP specimens), including in the near-threshold region. The higher thickness specimens have higher *da/dN* for the same Δ*K* values in all conditions analyzed in Figure 5. It is also possible to notice that, for *R* = 0.05, the increase in *da/dN* is higher with increasing Δ*K*. Specimen's thickness effect is more noticeable for lower *R*-values and *MP* samples. As expected, independently of the surface treatment, the thicker specimens have higher crack growth rates over all the Δ*K* range analyzed.

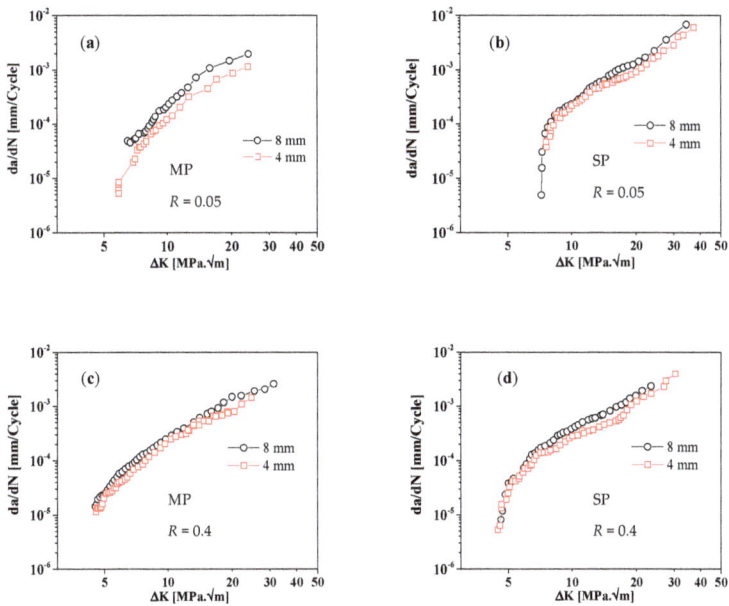

Figure 5. Thickness effect on the *da/dN-ΔK* curves for specimens: (**a**) MP, *R* = 0.05; (**b**) SP, *R* = 0.05; (**c**) MP, *R* = 0.4; and (**d**) SP, *R* = 0.4.

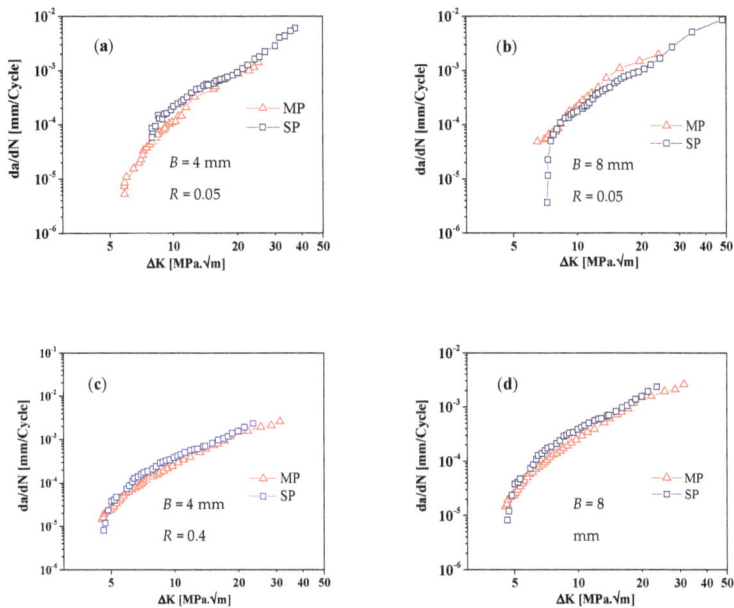

Figure 6. Shot peening effect on *da/dN-ΔK* curves for specimens: (**a**) *B* = 4 mm, *R* = 0.05; (**b**) *B* = 8 mm, *R* = 0.05; (**c**) *B* = 4 mm, *R* = 0.4; and (**d**) *B* = 8 mm, *R* = 0.4.

The main purpose of current work was the analysis of the surface peening effect on the crack propagation. Figure 6a–d shows the influence of the shot peening on the *da/dN-ΔK* curves for both thicknesses and stress ratios. Taking into account that shot peening has a very localized effect near the surface, which results in the introduction of compressive residual stresses, the propagation of cracks will be affected only in these areas. To analyze the retardation of crack propagation around the surface, the fractured specimens were observed by optical microscopy.

Figure 7a,b presents exemplary photos showing the marks of crack growth shape for machined and shot peening specimens, with 8 mm thickness, respectively. These marks were produced during variable amplitude loading with periodic overload blocks tests. Although the specimen's thickness is small to ensure tri-axial plain strain conditions in the center of the sample, the crack path presents a significant tunnel effect, as shown in Figure 7a,b, and also according to Zhou et al. [9]. The visual observation of the images does not show a clear evidence of the shot peening effect on the crack path. For a detailed analysis, a tunnel effect parameter was defined by the Equation (4):

$$tunnel\ effect = \frac{a_2 - \left(\frac{a_1 + a_3}{2}\right)}{a_0} \tag{4}$$

where a_1 and a_3 are the crack lengths at the specimen's surfaces, a_2 is the current crack length at the center and a_0 is the initial crack length. The tunnel effect is a well-studied manifestation in fatigue crack propagation. Specimens stress state affect fatigue crack propagation, thus propagation rate is distinct at the crack flanks front relatively to specimens' central points. The effect of stress state is usually explained by crack closure mechanisms. Typically, a plane stress state occurs at the surface that promotes crack tip plastic deformation and accordingly plasticity induces crack closure [32]. In turn, inside the specimen, there is a tri-axial stress state which prevents plastic deformation. As fracture surface roughness may be different, promoting roughness induces crack closure, especially for low values of ΔK [33]. This stress state effect on fatigue crack propagation slows down crack growth at the surface and hence promotes the tunnel effect. Several different parameters are used to understand to what extend tunnel affects the specimens' behavior. The simplest and most common parameter is *d/B* (Figure 7c). Other used parameters are used and presented in the literature [34,35]. Note that the stable shape of the crack front has a uniform distribution of effective stress intensity factor range.

The tunnel effect parameter is plotted in Figure 7d against the fatigue crack length (a_2–a_0). As expected, shot peening increases the retardation of the surface crack propagation observed by a higher tunnel effect parameter for crack length lesser than 10 mm. As mentioned above, tunnel effect can be caused by residual stresses profiles.

For MP specimens, average residual stress in load direction obtained from four measurements at the surface was about +290 MPa, while for SP samples compressive residual stresses occur around the surface. Figure 8 shows the profile of residual stresses and X-ray diffraction peak breadth against the depth from surface. According to the diffraction peak breadth profiles, the thickness layer affected by all surface treatments is circa 200 μm. The average value of the compressive residual stresses occurring trough a layer below the free surface with a 150 μm depth is about 200 MPa. Regarding the MP specimens, the residual stress measurements, for the same depth of the SP specimens presented an average value of 180 MPa. The reduced thickness of this layer, justifies the reduced influence on the overall propagation of fatigue cracks observed, in accordance to He et al. [36].

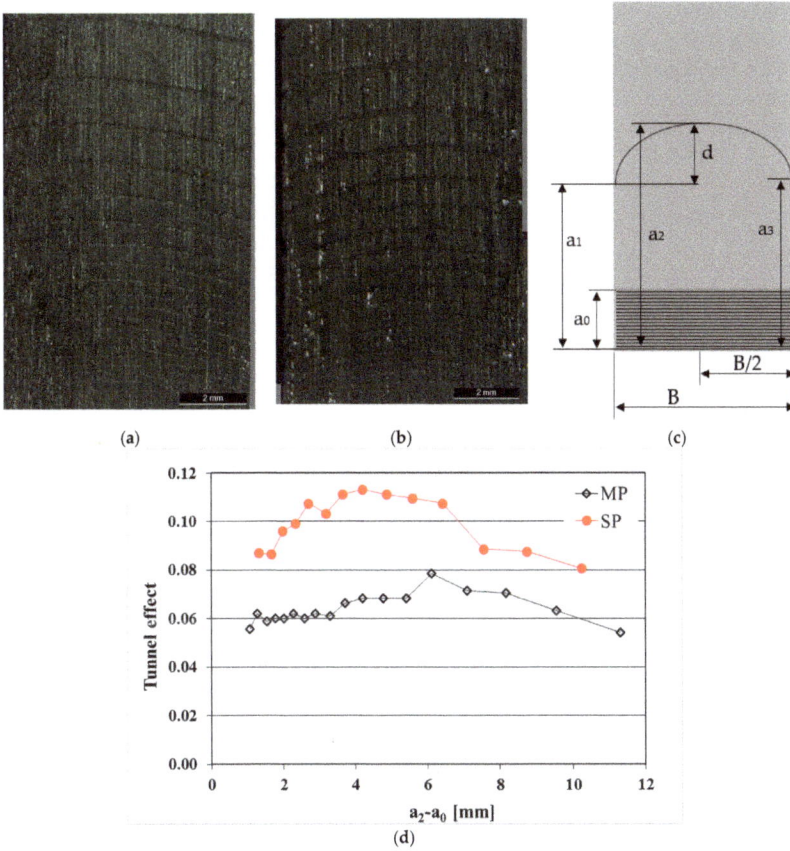

Figure 7. Impression marks of the crack path and tunnel effect: (**a**) MP specimens; (**b**) SP specimens; (**c**) schematic indication for tunnel effect parameters; and (**d**) tunnel effect value distribution for MP and SP specimens.

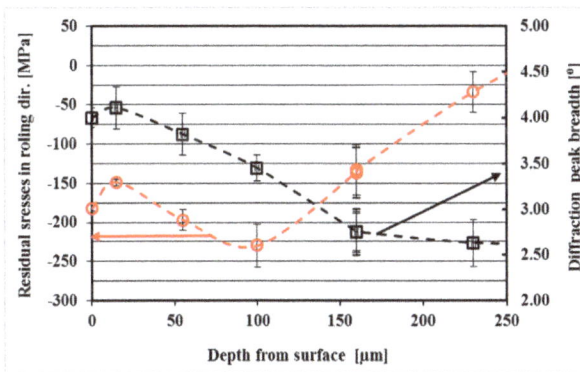

Figure 8. Residual stresses profile and X-ray diffraction peak breadth against the depth from surface for SP specimens. "○" corresponds to the residual stress in rolling direction (MPa) and "□" to the diffraction peak breadth (°).

Results obtained for both stress ratios $R = 0.05$ and $R = 0.4$ are compared in Figure 9, for both surface treatments and thicknesses. As expected, a meaningful influence of the stress ratio was noticed in both Paris law regime and near-threshold condition. As reported in the literature, this effect is mainly consequence of the significant reduction on crack closure level for higher stress ratio $R = 0.4$ [31]. This is why above 15 MPa√m the R does not present any significate effect.

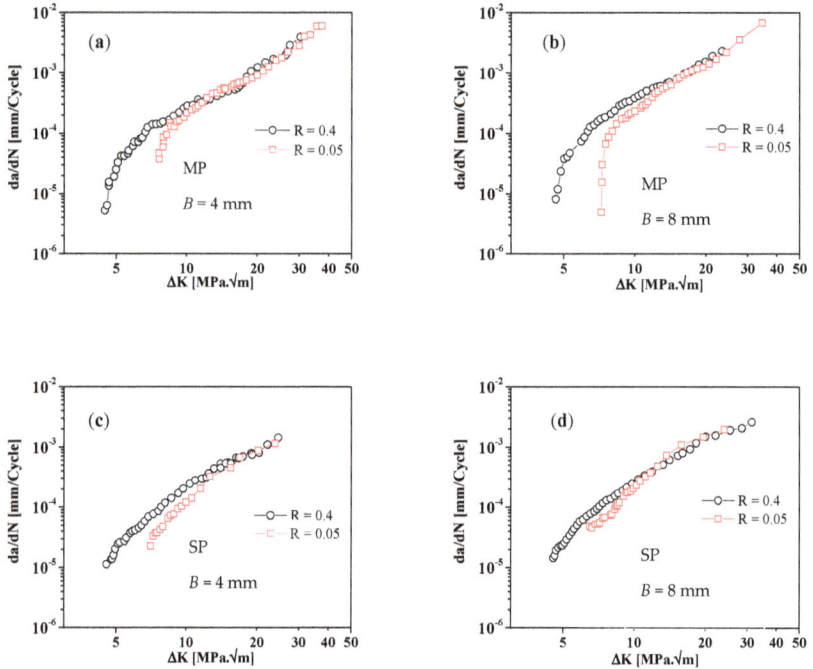

Figure 9. R effect on da/dN-ΔK curves for specimens: (**a**) $B = 4$ mm, MP; (**b**) $B = 8$ mm, MP; (**c**) $B = 4$ mm, SP; and (**d**) $B = 8$ mm, SP.

Fatigue crack propagation resulting in the stable regime were modeled by Paris law equation. Table 3 summarizes the values of the coefficients and intervals of validity of Paris' law and the correlation coefficients for all test conditions.

Table 3. Paris law parameters, C and m determined from da/dN vs. ΔK curves (mm/cycle; MPa m$^{1/2}$).

B [mm]	Specimen	R	C	m	Validity [MPa m$^{1/2}$]	Correlation Factor
4	MP	0.05	1.41×10^{-8}	3.94	7–13	0.995
4	MP	0.4	2.42×10^{-6}	2.04	12–24	0.996
4	SP	0.05	2.95×10^{-7}	2.94	8–14	0.970
4	SP	0.4	2.70×10^{-7}	3.05	5–10	0.982
8	MP	0.05	2.72×10^{-8}	3.89	7–12	0.996
8	MP	0.4	1.96×10^{-6}	2.16	13–22	0.998
8	SP	0.05	2.53×10^{-7}	2.97	9–16	0.991
8	SP	0.4	2.63×10^{-7}	3.25	5–17	0.973

To analyze the transient effects after overloads, variable amplitude loading with $R = 0.04$ were carried out, in which periodic overload blocks of 300 cycles were applied with intervals of N_{int} of 7500 and 15,000 cycles, as shown in Figure 3. The results obtained were compared with the reference constant amplitude loading tests. Figure 10a–d shows the collected results from the tests performed in specimens with 8 mm thick. The typical transient behavior after overloads is not detected in all blocks because of the reduced transient zone and the crack measuring method. The analysis of the figure shows that for MP specimens the fatigue crack growth rate reduction reaches the maximum value for N_{int} = 7500 cycles, while for the SP specimens the crack growth rate continues to decrease, although slightly, when N_{int} increases from 7500 to 15,000 cycles. For MP specimens, fatigue crack growth decreases more for 7500 cycles because induced plasticity of crack closure retardation is more critical (Figure 10d). This behavior cannot be confirmed when crack closure is not measured. For the SP specimens with R = 0, the behavior is similar. This effect is more noticeable for N_{int} = 7500 cycles then for 15,000 cycles.

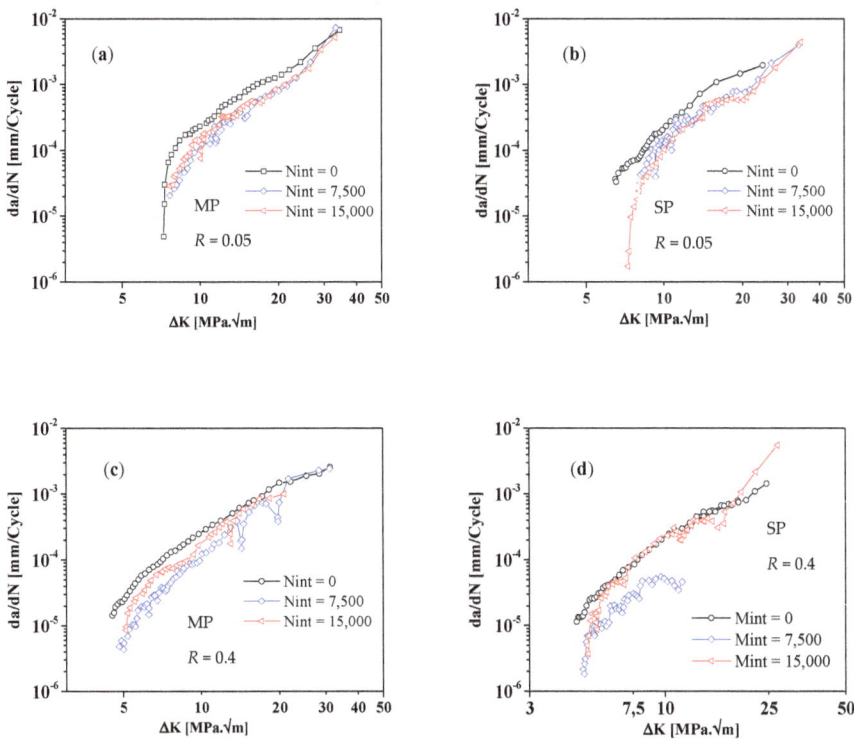

Figure 10. Block overload effect on *da/dN-ΔK* curves for 8 mm thick specimens: (**a**) R = 0.05, MP; (**b**) R = 0.05, SP; (**c**) R = 0.4, MP; and (**d**) R = 0.4, SP.

To understand better the fatigue mechanisms processes, fracture surfaces of the samples were observed in a Philips XL30 scanning electron microscope. Figure 11 shows two exemplary photos with different magnification of the crack propagated region in Paris' law regime, representative of various observations done during the study. Both images in Figure 11 show that fatigue crack propagation occurs mainly by striation.

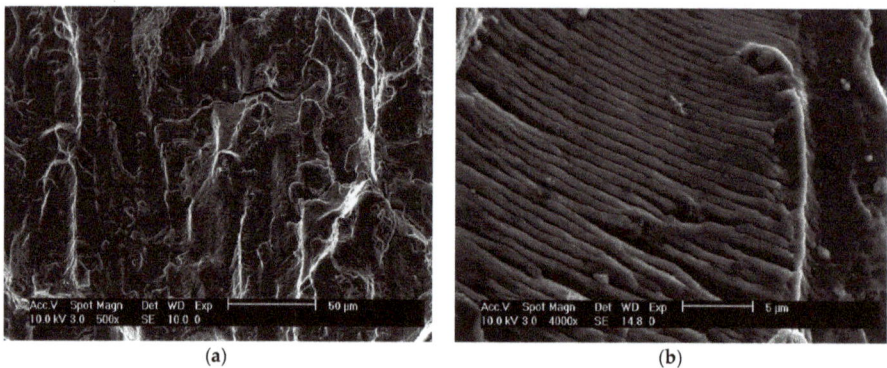

(a)   (b)

Figure 11. Exemplary fracture surface morphology from SEM observations.

4. Conclusions

The present work studied the effects of the shot peening and the stress ratio on the fatigue crack propagation of the 7475 aluminum alloy with a T7351 heat treatment, using two specimens' thickness: 4 and 8 mm. The analysis of the results draws the following conclusions:

- As a result of its small influence depth, the beneficial effect of shot peening on da/dN-ΔK curves is negligible, particularly for $R = 0.4$. However, this effect seems to increase near the threshold condition.
- For both mechanically polished and shot-peened samples, a specimen's thickness has only marginal influence on the stable crack propagation regime.
- A significant effect of the mean stress was observed, particularly in near- threshold region.
- Periodic overload blocks promote a reduction of the fatigue crack growth rate. For MP specimens, the reduction reaches the maximum value for the interval between blocks of 7500 cycles, while, for SP specimens, the crack growth rate continues to decrease for intervals of 15,000 cycles.

Acknowledgments: This research is sponsored by FEDER funds through the program COMPETE—Programa Operacional Factores de Competitividade—and by national funds through FCT—Fundação para a Ciência e a Tecnologia—under the project PEst-C/EME/UI0285/2013. The authors would also like to acknowledge OGMA-Indústria Aeronáutica de Portugal, Alverca, Portugal, and Dra Ana Guimarães and Eng. João Miranda for the collaboration in the shot-peening processing.

Author Contributions: José A. M. Ferreira. and José D. M. Costa conceived and designed the experiments; Pedro V. Antunes, Natália Ferreira, and Carlos Capela performed the experiments; Pedro V. Antunes and Carlos Capela analyzed the data; Carlos Capela contributed with the machining of test specimens; Pedro V. Antunes, José A. M. Ferreira, and José D. M. Costa wrote the paper.

Conflicts of Interest: The authors declare no conflict of interest.

References

1. Fathallah, R.; Laamouri, A.; Sidhom, H.; Braham, C. High cycle fatigue behavior prediction of shot-peened parts. *Int. J. Fatigue* **2004**, *26*, 1053–1067. [CrossRef]
2. Miková, K.; Bagherifard, S.; Bokůvka, O.; Guagliano, M.; Trško, L. Fatigue behavior of X70 microalloyed steel after severe shot peening. *Int. J. Fatigue* **2013**, *55*, 33–42. [CrossRef]
3. Bagherifard, S.; Guagliano, M. Fatigue behavior of a low alloy steel with nanostructured surface obtained by severe shot peening. *Eng. Fract. Mech.* **2012**, *81*, 56–68. [CrossRef]
4. Zhang, P.; Lindemann, J.; Leyens, L. Shot peening on the high-strength magnesium alloy AZ80—effect of peening media. *J. Mater. Process. Technol.* **2010**, *210*, 445–450. [CrossRef]

5. Petit, J.; Mendez, J. Some aspects of the influence of microstructure on fatigue resistance. In *Fatigue 96, Proceedings of the Sixth International Fatigue Congress, Berlin, Germany, 6–10 May 1996*; Lutjering, G., Nowack, H., Eds.; Pergamon: Oxford, UK, 1996; Volume I, pp. 15–26.

6. Bergner, F.; Zouhar, G. A new approach to the correlation between the coefficient and the exponent in the power law equation of fatigue crack growth. *Int. J. Fatigue* **2000**, *22*, 229–239. [CrossRef]

7. Bergner, F.; Zouhar, G.; Tempus, G. The material-dependent variability of fatigue crack growth rates of aluminium alloys in the Paris regime. *Int. J. Fatigue* **2001**, *23*, 383–394. [CrossRef]

8. Sunder, R.; Dash, P.K. Measurement of fatigue crack closure through electron microscopy. *Int. J. Fatigue* **1982**, *4*, 97–105. [CrossRef]

9. Zhou, J.Z.; Huang, S.; Sheng, J.; Lu, J.Z.; Wang, C.D.; Chen, K.M.; Ruan, H.Y.; Chen, H.S. Effect of repeated impacts on mechanical properties and fatigue fracture morphologies of 6061-T6 aluminium subject to laser peening. *Mater. Sci. Eng. A* **2012**, *539*, 360–368. [CrossRef]

10. Pant, B.K.; Pavan, A.H.V.; Prakash, R.V.; Kamaraj, M. Effect of laser peening and shot peening on fatigue striations during FCGR study of Ti6Al4V. *Int. J. Fatigue* **2016**, *93*, 38–50. [CrossRef]

11. Vecchio, R.S.; Hertzberg, R.W.; Jaccard, R. On the overload induced fatigue crack propagation behavior in aluminium and steel alloys. *Fatigue Fract. Eng. Mater. Struct.* **1984**, *7*, 181–194. [CrossRef]

12. Ward-Close, C.M.; Blom, A.F.; Richie, R.O. Mechanisms associated with transient fatigue crack growth under variable amplitude loading: An experimental and numerical study. *Eng. Fract. Mech.* **1989**, *32*, 613–638. [CrossRef]

13. Krumar, R.; Garg, S.B.L. Effect of yield strength and single overload cycles on effective stress intensity range ratio in 6061-T6 Alalloy. *Eng. Fract. Mech.* **1989**, *34*, 403–412. [CrossRef]

14. Ling, M.R.; Schijve, J. The effect of intermediate heat treatments on overload induced retardations during fatigue crack growth in an Al-alloy. *Fatigue Fract. Eng. Mater. Struct.* **1992**, *15*, 421–430. [CrossRef]

15. Damri, D.; Knott, J.F. Fracture modes encountered following the application of a major tensile overload cycle. *Int. J. Fatigue* **1993**, *15*, 53–60. [CrossRef]

16. Shuter, D.M.; Geary, W. Some aspects of fatigue crack growth retardation behaviour following tensile overloads in a structural steel. *Fatigue Fract. Eng. Mater. Struct.* **1996**, *19*, 185–199. [CrossRef]

17. Robin, C.; Louah, M.; Pluvinage, G. Influence of the overload on the fatigue crack growth in steels. *Fatigue Fract. Eng. Mater. Struct.* **1983**, *6*, 1–13. [CrossRef]

18. Shercliff, H.R.; Fleck, N.A. Effect of specimen geometry on fatigue crack growth in plane strain—II. Overload response. *Fatigue Fract. Eng. Mater. Struct.* **1990**, *13*, 297–310. [CrossRef]

19. Shin, C.S.; Hsu, S.H. On the mechanisms and behaviour of overload retardation in AISI 304 stainless steel. *Int. J. Fatigue* **1993**, *15*, 181–192. [CrossRef]

20. Dexter, R.J.; Hudak, S.J.; Davidson, D.L. Modelling and measurement of crack closure and crack growth following overloads and underloads. *Eng. Fract. Mech.* **1989**, *33*, 855–870. [CrossRef]

21. Tsukuda, H.; Ogiyama, H.; Shiraishi, T. Transient fatigue crack growth behaviour following single overloads at high stress ratios. *Fatigue Fract. Eng. Mater. Struct.* **1996**, *19*, 879–891. [CrossRef]

22. Borrego, L.P.; Ferreira, J.M.; Pinho da Cruz, J.M.; Costa, J.M. Evaluation of overload effects on fatigue crack growth and closure. *Eng. Fract. Mech.* **2003**, *70*, 1379–1397. [CrossRef]

23. Donald, K.; Paris, P.C. An evaluation of DKeff estimation procedures on 6061-T6 and 2024-T3 aluminium alloys. *Int. J. Fatigue* **1999**, *21*, S47–S57. [CrossRef]

24. Paris, P.C.; Tada, H.; Donald, J.K. Service load fatigue damage—-A historical perspective. *Int. J. Fatigue* **1999**, *21*, S35–S46. [CrossRef]

25. Borrego, L.P.; Ferreira, J.M.; Costa, J.M. Fatigue crack growth and crack closure in an AlMgSi alloy. *Fatigue Fract. Eng. Mater. Struct.* **2001**, *24*, 255–265. [CrossRef]

26. The American Standard SAE J443. *Procedures for Using Standard Shot Peening Test Strip*; American Standard: Piscataway Township, NJ, USA, 1968.

27. The American Standard. *Aerospace Materials Division*; SAE, Shot Peening, SAE AMS 2430; American Standard: Piscataway Township, NJ, USA, 2009.

28. American Society for Testing and Materials. Standard Test Method for Microhardness of Materials. In *Annual Book of ASTM Standards*; ASTM: West Conshohocken, PA, USA, 2000; Volume 03.01, p. E647.

29. International Organization for Standardization. *DIN EN ISO 4288: Geometrical Product Specifications (GPS). Surface Texture: Profile Method: Rules and Procedures for the Assessment of Surface Texture*; ISO: Geneva, Switzerland, 1996.

30. American Society for Testing and Materials. *Standard Test Method for Vickers Indentation Hardness of Advanced Ceramics*; ASTM: West Conshohocken, PA, USA, 2015; p. C1327.
31. Borrego, L.P.; Costa, J.D.M.; Silva, S.; Ferreira, J.M. Microstructure dependent fatigue crack growth in aged hardened aluminium alloys. *Int. J. Fatigue* **2004**, *26*, 1321–1331. [CrossRef]
32. Antunes, F.V.; Chegini, A.G.; Branco, R.; Camas, D. A numerical study of plasticity induced crack closure under plane strain conditions. *Int. J. Fatigue* **2015**, *71*, 75–86. [CrossRef]
33. Antunes, F.V.; Ramalho, A.L.; Ferreira, J.A.M. Identification of Fatigue Crack Propagation Modes with Roughness Measurements. *Int. J. Fatigue* **2000**, *22*, 781–788. [CrossRef]
34. Lin, X.B.; Smith, R.A. Fatigue shape analysis for corner cracks at fastener holes. *Eng. Fract. Mech.* **1998**, *59*, 73–87. [CrossRef]
35. Branco, R. Numerical Study of Fatigue Crack Growth in MT Specimens. Master's Thesis, Department of Mechanical Engineering, University of Coimbra, Coimbra, Portugal, 2006.
36. He, B.Y.; Soady, K.A.; Mellor, B.G.; Harrison, G.; Reed, P.A.S. Fatigue crack growth behaviour in the LCF regime in a shot peened steam turbine blade material. *Int. J. Fatigue* **2016**, *82*, 280–291. [CrossRef]

applied sciences

MDPI

Article

Improving Stability Prediction in Peripheral Milling of Al7075T6

Daniel Olvera [1], Gorka Urbikain [2,*], Alex Elías-Zuñiga [1] and Luis Norberto López de Lacalle [2]

[1] Tecnológico de Monterrey, Escuela de Ingeniería y Ciencias, Av. Eugenio Garza Sada 2501, Monterrey, Nuevo León 64849, Mexico; daniel.olvera.trejo@itesm.mx (D.O.); aelias@itesm.mx (A.E.-Z.)
[2] Department of Mechanical Engineering, University of the Basque Country, Alameda de Urquijo s/n, 48013 Bilbao, Bizkaia, Spain; norberto.lzlacalle@ehu.es
* Correspondence: gorka.urbikain@ehu.es; Tel.: +34-943-018-643

Received: 11 July 2018; Accepted: 1 August 2018; Published: 7 August 2018

Abstract: Chatter is an old enemy to machinists but, even today, is far from being defeated. Current requirements around aerospace components call for stronger and thinner workpieces which are more prone to vibrations. This study presents the stability analysis for a single degree of freedom down-milling operation in a thin-walled workpiece. The stability charts were computed by means of the enhanced multistage homotopy perturbation (EMHP) method, which includes the helix angle but also, most importantly, the runout and cutting speed effects. Our experimental validation shows the importance of this kind of analysis through a comparison with a common analysis without them, especially when machining aluminum alloys. The proposed analysis demands more computation time, since it includes the calculation of cutting forces for each combination of axial depth of cut and spindle speed. This EMHP algorithm is compared with the semi-discretization, Chebyshev collocation, and full-discretization methods in terms of convergence and computation efficiency, and ultimately proves to be the most efficient method among the ones studied.

Keywords: numerical methods; milling; computation; stability

1. Introduction

Chatter is a dynamic instability phenomenon that diminishes the quality of parts and tool performance. It is an old adversary for machinists, and one of the most common issues in manufacturing. However, nowadays, due to the rapid growth of global competition to reduce cost and the increased dimensional accuracy in monolithic and thin geometries demanded by aeronautical industries, research has been focused on more accurate predictive models and methods for optimizing metal removal rates without chatter. Those models are delay-differential equations (DDE) with infinite dimensional state space. Particularly in milling, the equation of motiving is a DDE with a time-periodic coefficient, transformed in a finite dimensional system. According to the Floquet theory, the stability properties are determined by the system's monodromic operator.

The literature provides various approaches for predicting stability lobes in various machining operations as a function of spindle speed and chip load. For instance, Smith and Tlusty et al. [1] presented a method for generating the stability lobes in milling operations by using time-domain numerical simulations. Altintas and Budak [2] developed the first analytical solution that led to the milling operation prediction of stability lobes in the frequency domain by averaging the time-dependent periodic directional coefficients. This method provides accurate stability predictions. except for where there are cutting operations at low radial immersion. Davies et al. [3] proved that the traditional regenerative stability theory could not accurately predict stability lobes at low radial immersions. However, Merdol and Altintas [4] proved that the multi-frequency approach [5] was able to accurately predict low radial immersion milling stability lobes if the harmonics of the tooth-passing frequencies were included in the eigenvalue solution. The double-period bifurcation at small radial immersion was investigated by Bayly et al. [6] by applying the time finite element analysis (TFEA) in the solution of the governing equations of motion. The approximate solution obtained from this method was cast in the form of a discrete map that related position and velocity at the beginning and end of each element. Then, the eigenvalues of the discrete map were used to determine the stability bounds. Another method that forms a finite dimensional transition matrix as an approximation of the infinite dimensional monodromic operator is the semi-discretization (SD) method. In this method, first introduced by Insperger and Stepan [7], the delayed terms are discretized, the undelayed terms unchanged, and the time-periodic coefficients approximated by piecewise constant functions. An updated version of the SD method for periodic systems with a single discrete time delay was proposed by Insperger and Stepan [8]. In this approach, the time step is chosen as an integer fraction of the time period, meaning that the Floquet transition matrix is determined over a single period. Insperger et al. showed that the second and higher-order approximations of the delayed term did not provide better convergence than the first-order one [9,10]. On the other hand, Butcher et al. [11] developed a new approximation technique for studying the stability properties of milling operations. This approach was based on the properties of Chebyshev polynomials and used a collocation representation of the solution at their endpoints. The stability bounds of the corresponding equations were determined by the eigenvalues of the approximate monodromic matrix, mapping function values at the collocation points from one interval to the next. They concluded that the Chebyshev collocation (CH) approach provided results similar to other methods which used equispaced points. This approach was extended by the authors in complex turning applications [12,13] and compared to other computation techniques [14].

Ding et al. [15] developed a full-discretization (FD) method based on the direct integration scheme for prediction of milling stability lobes by taking the regenerative effect into account in the state-space which is represented in integral form. They claimed that this method had better computational efficiency than the SD method developed by Insperger and Stepan [8]. However, Insperger [16] showed that the full-discretization and SD methods were similar, as both methods approximated the delay-differential equations by using a series of ordinary differential equations, meaning that the FD method was an alternative to the SD method with only a slightly different concept in the discretization scheme.

Recently, Ding et al. [17] proposed a second-order full-discretization method to determine the stability lobes of milling operations. After the time period was equally discretized into a finite set of intervals, the full-discretization method was developed to handle the integration term of the system. They showed the practicality of the second-order full-discretization method, in terms of both accuracy and computational time efficiency. The authors [18] adapted the homotopy theory to predict stability lobes by proposing an initial solution and then deforming it to accurately approximate the monodromic operator by EMHP method. Since there are various approaches in literature to predict stability properties of DDEs, it is important to investigate the capabilities of each method in terms of convergence and computation time over a more complicated modeling of forces in the milling process.

2. Materials and Methods

2.1. Milling Stability Prediction with the EMHP Method

In order to predict stability in machining operations, in this paper we focus on the manufacturing of a thin-walled workpiece through peripheral milling operations. The workpiece has a single degree of freedom (flexibility only in the y-direction according to the machine tool standard axis nomenclature), with a dynamic model of the following form:

$$\ddot{y}(t) + 2\zeta\omega_n\dot{y}(t) + \omega^2{}_n y(t) = \frac{F_y}{m_m}(y(t) - y(t - \tau)) \tag{1}$$

Here, m_m represents the modal mass, ζ is the damping ratio, ω_n is the natural angular frequency, and F_y is the cutting force over the workpiece in the y-direction, which is given by:

$$F_y(t) = \sum_{j=1}^{z_n} g\left(\phi_j(t)\right) \cos\left(\phi_j(t)\right)\left(-K_{tc} \sin\left(\left(\phi_j(t)\right)\right) + K_{rc} \cos\left(\left(\phi_j(t)\right)\right)\right) \tag{2}$$

where z_n is the number of teeth, K_{tc} and K_{rc} are the tangential and the normal cutting force coefficients, and ϕ_j *(t)* is defined as:

$$\phi_j(t) = (2\pi n/60)t + j2\pi/z_n \tag{3}$$

where n is the spindle speed in rpm. The function $g(\phi_j(t))$ is a screen function, and is equal to 1 if the tooth j is in the cut, and is equal to 0 if j is out of the cut. To obtain the stability lobes of Equation (1), we proposed a solution procedure based on the EMHP algorithm [18], which relies on a sequence of subintervals that provide approximate solutions requiring less CPU time in comparison with other methods found in the literature. In this methodology, we rewrote Equation (1) as:

$$\ddot{y}_i(T) + 2\zeta\omega_n\dot{y}_i(T) + \omega_n y_i(T) \approx \frac{F_{yt}}{m_m}\left(y_i(T) - y_i^T(T)\right) \tag{4}$$

where $F_{yt} = F_y(t)$, and $y_i(T)$ denotes the approximate solution of order m in the i-th sub-interval that satisfies the initial conditions $y_i(0) = y_{i-1}$, $\dot{y}_i(0) = \dot{y}_{i-1}$. The Equation (1) can be written in state space, in accordance with the following matrix-form representation:

$$\dot{x}(t) = \mathbf{A}(t)\mathbf{x}(t) + \mathbf{B}(t)(\mathbf{x}(t) - \mathbf{x}(t - \tau)) \tag{5}$$

where $x = [x, \dot{x}]^T$, $\mathbf{A}(t - \tau) = \mathbf{A}(t)$, $\mathbf{B}(t - \tau) = \mathbf{B}(t)$, and τ is the time delay. By following the homotopic procedure, the equivalent form of Equation (5) can be written as:

$$\dot{x}_i(T) - \mathbf{A}_t\mathbf{x}_i(T) \approx \mathbf{B}_t\mathbf{x}_i^T(T) \tag{6}$$

where $x_i(T)$ denotes the m order solution of Equation (6) in the i-th sub-interval that satisfies the initial conditions $\mathbf{x}_i(0) = \mathbf{x}_{i-1}$, $\mathbf{A}_t$, and $\mathbf{B}_t$ represents the values of the periodic coefficients at the time t. In order to approximate the delay term $\mathbf{x}_i^T(T)$ in Equation (6), the period $[t_0 - t_0]$ is discretized into N equally-spaced points, but the method does not accept strictly-spaced sub-intervals. Thus, each sub-interval has a time span equal to $\Delta t = \tau/(N - 1)$. Here, we assumed that the function $\mathbf{x}_i^T(T)$ in the delay sub-interval $[t_{i-N}, t_{i-N+1}]$ had a first-order polynomial representation, of the form:

$$\mathbf{x}_i^T(T) = \mathbf{x}_{i-N+1}(T) \approx \mathbf{x}_{i-N} + \frac{N-1}{\tau}(\mathbf{x}_{i-N+1} - \mathbf{x}_{i-N})T \tag{7}$$

To simplify the notation, we let $\mathbf{x}_i \equiv \mathbf{x}_i(T_i)$. Equation (7) was then introduced into Equation (6), to get:

$$\dot{x}_i(T) = \mathbf{A}_t\mathbf{x}_i(T) + \mathbf{B}_t\mathbf{x}_{i-N} - \frac{N-1}{\tau}\mathbf{B}_t\mathbf{x}_{i-N}T + \frac{N-1}{\tau}\mathbf{B}_t\mathbf{x}_{i-N+1}T \tag{8}$$

where:

$$\mathbf{A}_t = \begin{bmatrix} 0 & 1 \\ -\omega_n^2 + \frac{F_{yt}}{m_m} & -2\zeta\omega_n \end{bmatrix}, \; \mathbf{B}_t = \begin{bmatrix} 0 & 0 \\ -\frac{F_{yt}}{m_m} & 0 \end{bmatrix} \tag{9}$$

By following our homotopic procedure, we assumed the homotopic representation of Equation (8) was of the form:

$$H(\mathbf{X}_i, p) = L(\mathbf{X}_i) - L(\mathbf{x}_{i0}) + pL(\mathbf{x}_{i0}) = p\left(\mathbf{AX}_i + \mathbf{Bx}_{i-N} - \frac{N-1}{\tau}\mathbf{Bx}_{i-N}T + \frac{N-1}{\tau}\mathbf{Bx}_{i-N+1}T\right) \tag{10}$$

To obtain the stability lobes of Equation (1), the solution of Equation (10) was rewritten in the form:

$$\mathbf{x}_i(T) \approx \mathbf{P}_i(T)\mathbf{x}_{i-1} + \mathbf{Q}_i(T)\mathbf{x}_{i-N} + \mathbf{R}_i(T)\mathbf{x}_{i-N+1} \tag{11}$$

where:

$$\mathbf{P}_i(T) = \sum_{k=0}^{m} \frac{1}{k!}\mathbf{A}_t^k T^k,$$

$$\mathbf{Q}_i(T) = \begin{cases} \sum_{k=1}^{m} \frac{N-1}{(k+1)!(\tau)} \mathbf{A}_t^{k-1}\mathbf{B}_t T^{k+1} & m \geq 1 \\ 0 & m = 0 \end{cases} \tag{12}$$

$$\mathbf{R}_i(T) = \begin{cases} \sum_{k=1}^{m} \frac{1}{k!}\mathbf{A}_t^{k-1}\mathbf{B}_t T^k - \mathbf{Q}_m & m \geq 1 \\ 0 & m = 0 \end{cases}$$

The approximate solution given by Equation (12) can be written as a discrete map by using the following identity:

$$\mathbf{w}_i = \mathbf{D}_i\mathbf{w}_{i-1} \tag{13}$$

where $\mathbf{D}_i$ is a coefficient matrix, w_{i-1} is a vector of the form:

$$\mathbf{w}_{i-1} = \left[x_{i-1}, \dot{x}_{i-1}, x_{i-2}, \ldots, x_{i-N}\right]^T \tag{14}$$

By using Equations (11) and (12), we could easily construct the coefficient matrix $\mathbf{D}_i$ [8]. We then determined the transition matrix $\boldsymbol{\Phi}$ over the period $\tau = (N-1)\Delta t$ by coupling each approximate solution through the discrete map $\mathbf{D}_i$, $i = 1, 2, \ldots, (N-1)$, to get:

$$\boldsymbol{\Phi} = \mathbf{D}_{N-1}\mathbf{D}_{N-2}\ldots\mathbf{D}_2\mathbf{D}_1 \tag{15}$$

Then, the stability lobes of Equation (1) were determined by computing the eigenvalues of the transition matrix given by Equation (15).

2.2. Effect of Runout and Cutting Speed

In this section, the effects of runout and cutting speed in the peripheral milling of Al7075T6 alloy at high cutting speeds are investigated—or more specifically, their influence in the prediction and accuracy of stability lobes is analyzed.

Because perfect cutting is impossible, runout appears naturally in every milling operation and unfortunately produces undesirable effects, such as uneven material removal, wear, and poor surface quality. It is defined as an offset of the tool's geometrical axis with respect to the spindle rotation axis, as shown in Figure 1. This is impossible to avoid because the total runout is a sum of the misalignments of the spindle, tool-holder, and milling tool itself. This makes difficult to predict the current flutes' trajectories, as the real cheap thickness is defined by the trajectory at every point in each flute, which changes not only in the *x*-*y* plane but also in space.

When considering this effect during milling operations, the cutting force can be expressed as:

$$F_y(t) = \sum_{j=1}^{z_n} \rho_j \frac{1}{4k_\beta} \left[K_{tc}\cos 2\phi_j(z) + K_{rc}\left(\sin 2\phi_j(z) + 2\phi_j(z)\right)\right]_{z_{j,1}(\phi_j)}^{z_{j,2}(\phi_j)} \tag{16}$$

where ρ_j represents a weighted factor that is related to the cutting force magnitudes for each flute, *j*. Here, we propose a different alternative to consider the runout effects, by assuming that relating ρ_j is given as:

$$\rho_j = \left[1 + \delta\left(\cos\phi_p(j-1) - \cos\phi_p(j-2)\right)\right], \ \delta = \frac{\tilde{r}}{f_z} \tag{17}$$

where $\tilde{r}$ represents the tool's apparent runout radius in the feed direction, α is the runout angle, and f_z is the feed rate. Notice that δ represents the chip variation in the feed direction (x-axis). To determine the value of $\tilde{r}$ experimentally, once the milling tool is attached to in the machine center, a deflection gauge fixed to the table was set to zero once the first flute pushed the gauge. Then, the apparent radius $\tilde{r}$ was estimated by the average deflection between 180°-spaced flutes. The proportional factor ρ_j for the chip load in the actual j teeth was then calculated as the nominal feed rate plus the ratio $\delta = \tilde{r}/f_z$, representing the percentage of the nominal feed rate.

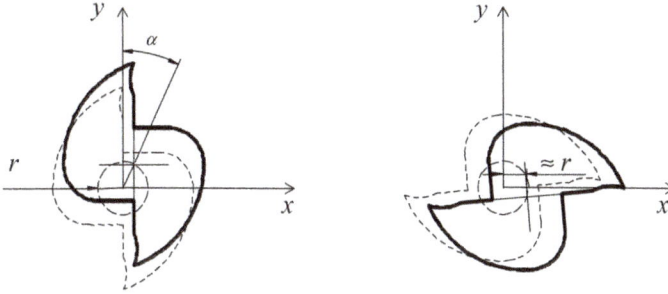

Figure 1. (**Left**) geometrical runout; (**Right**) approximated runout.

Note that the factor δ is related to the amount of cutting force magnitude of each tooth that depends on the apparent runout radius. In the ideal case of no runout, i.e., $\delta = 0$ and $\rho_j = 1$, Equation (15) becomes periodic in τ. On the other hand, when the runout influences the machining processes, the periodic behavior arises, related to the spindle period τ_T.

When the runout is zero, the dynamical system exhibits only quasi-periodic (Hopf bifurcation) and flip-bifurcation behaviors of period two, while chatter in period one never arises. In multi-flute machining processes, the runout practically always arises since the tool is not perfectly symmetric. In this case, the principal period is equal to the spindle rotation period $\tau_T = 60/n$. Thus, the determination of the stability lobes by the EMHP method can be done by computing the transition matrix Φ over the spindle rotation period, i.e., by considering that Equation (15) can be written as:

$$\Phi = D_{z_n(N-1)} D_{n_z(N-1)-1} \cdots D_2 D_1 \qquad (18)$$

Within this study, a 2-fluted, 16 mm-diameter, carbide end-mill (30° helix) was used for the experiments. The apparent runout was measured by means of a digital gauge, and the apparent radius was estimated to be at 17 μm. In order to validate the proposed runout model, this was compared with the chip thickness in the time domain by a time domain simulation similar to that developed in [19]. The results of this are shown in Figure 2. Note that the model matches the maximum thickness in every single flute, but it does not capture the discontinuity when one of the flutes is not cutting. In spite of this, the analytical model is a good approximation for the stability analysis.

When considering runout effects, new chatter frequencies appeared due to the one-period bifurcation, which was many times the spindle rotation frequency of a two-flute end mill (see Figure 3). The plotted chatter frequencies were also used when comparing with experimental chatter frequencies for validation later on.

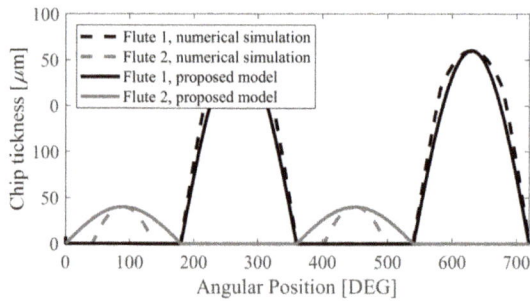

Figure 2. Proposed runout model versus a time domain simulation runout.

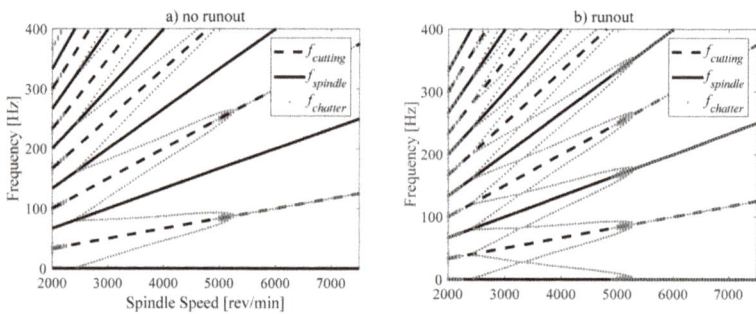

Figure 3. Chatter frequencies, (**a**) no runout, and (**b**) runout.

With regard to cutting speed, it is commonly assumed in the literature that the stability analysis and mechanistic force models are independent of this parameter [20,21]. However, during the machining of aluminum workpieces, cutting speed has a strong influence on the magnitude of the cutting forces and on the stabilities' boundaries. According to Asthakov [22], cutting speed affects the chip formation process in two major ways. Firstly, it changes the strain rate in the deformation zone, which influences the resistance of the workpiece material and the thermal energy generated during its deformation processes. Secondly, the cutting speed influences the tool-chip's relative speed and the natural length of the tool-chip interface.

Therefore, a proper characterization should consider variable cutting coefficients. In this work, cutting coefficients in the tangential and normal directions were obtained as functions of the cutting speed. Up to 120 full-immersion cutting tests were done by using tools with 1 mm–4 mm depth of cut, 5 feed rates (0.05–0.25 mm), and 6 cutting velocities (2000–10,000 rpm), as shown in Figure 4. These tests were performed on a Kistler dynamometer (9255 B). The solid end-mill was a 30° helix, which was 16 mm in diameter with two flutes, and the material workpiece selected was aluminum 7075-T6. It is important to mention that the associated spindle speed was not greater than 1/5 of the natural frequency of the chain table dynamometer workpiece ($\approx$1.6 KHz) in order to obtain reliable measurements [23].

During the interpretation of the experimental data, shear cutting coefficients showed significant variations due to the cutting speed in the radial and tangential directions. However, the fact that friction effects contributed to the total magnitude of the cutting forces due to cutting speed is negligible. We used a simple exponential model to fit the experimental data:

$$K_{qc}(v_c) = 4\left(A_q e^{-\alpha_q v_c} + B_q\right) \tag{19}$$

where the parameters for the fitting are A_q, B_q, and α_q are obtained for the tangential and normal directions, as: $A_t = 53.3$, $B_t = 201$, $\alpha_t = 0.0067$, $A_r = 84.4$, $B_t = 20$, and $\alpha_t = 0.0045$.

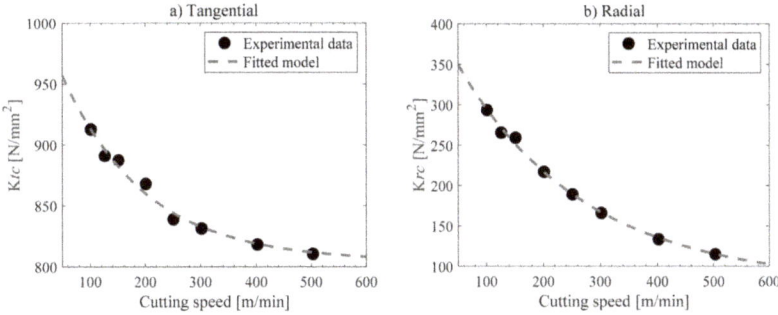

Figure 4. Experimental and fitted curves of cutting shear coefficients in the (**a**) tangential and (**b**) radial directions.

3. Results

3.1. Experimental Validation of Stability Prediction

In order to validate the proposed approach, down-milling tests were conducted in a conventional milling center. A monolithic artifact of Al7075T6 was designed to reproduce the dynamic response of a single degree of freedom system in the *y*-axis, which is orthogonal to the feedrate direction. The workpiece displacements were recorded and analyzed via Fast Fourier Transform, to observe the frequency content and to compare it with tool cutting frequencies (see Figure 3).

The modal parameter values of the end-milling cutting processes were found by impact testing: $\omega_n = 499$ rad/s, $\zeta = 0.04$, and $m_m = 10.1$ kg. The transfer function obtained through the impact test fit well with the theoretical results. By using these values and computing the eigenvalues of the transition matrix, the stability lobes of the end-milling cutting operation were calculated. Figure 5 shows the stability charts for down-milling with 25% radial immersion, feed rate 0.05 mm/teeth, and a measured apparent radius of 17 μm. The discontinued lines correspond to the typical stability lobes when the cutting speed, runout, and helix angle are neglected, while the solid lines plot the stability boundaries, taking these effects into account.

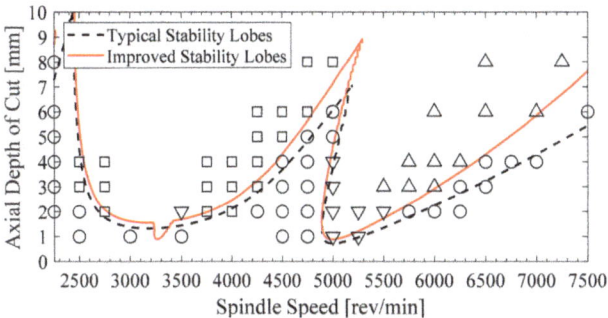

Figure 5. Experimental data from the down-milling operation, comparison between typical and improved stability analysis.

The stable cutting data are denoted by circles, unstable quasi-periodic by square symbols, and unstable double periodic by diamonds. Down triangles denote one period chatter respectively, and up triangles denote quasi-one-period data. When it was difficult to distinguish between experiments with stable and unstable data due to there not being a dominant frequency, the vibration amplitude was considered as a second criterion. The experimental data demonstrated the high accuracy of the stability boundary prediction, even to predict the transition between quasi-periodic and double-periodic unstable zones. For comparison, the effects of the runout and cutting speed were also introduced in the CH, SD, FD methods. In this case, the boundary limits obtained were almost identical when $N = 30$, however, the convergence rate and computation times differed depending on the method.

3.2. Accuracy and Computation Times

Since the convergence of the eigenvalues depends on parametric values (n, a_p), a more general approach has been studied using an 11×11 matrix containing an equally-spaced sweep of spindle speeds (2000 rpm to 7500 rpm) and axial depths of cut (0 to 10 mm). The difference between the matrix's approximate critical eigenvalues μ and exact ones μ_0 is presented as the function of the number of discretizations, or subintervals. The exact matrix of the eigenvalues μ_0 was determined by each corresponding method with $N = 200$. Figure 6 shows the results for 25% radial immersion. Note that this figure does not compare the converged values between methods, as no method converges at the same eigenvalues.

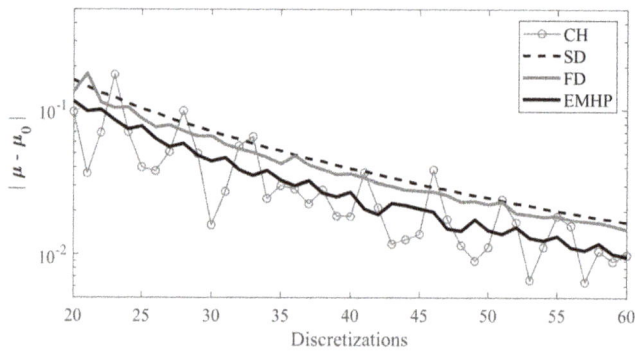

Figure 6. Convergence of the norm of the 11×11 matrix of eigenvalues for all methods.

Another important parameter is the time it takes to compute the stability lobes. Here, the average time needed to compute an eigenvalue for a combination of parameters (n, a_p) is used, which was calculated over the 11×11 matrix used previously. Figure 7 shows the average time for each method as a function of the number of discretization or sub-intervals. Once again, it is clear that the fastest methods for the stability performed were the CH and EMHP methods. It is important to note that the immersion does not change the number of mathematical operations. Figure 7 is valid for any value of radial immersion.

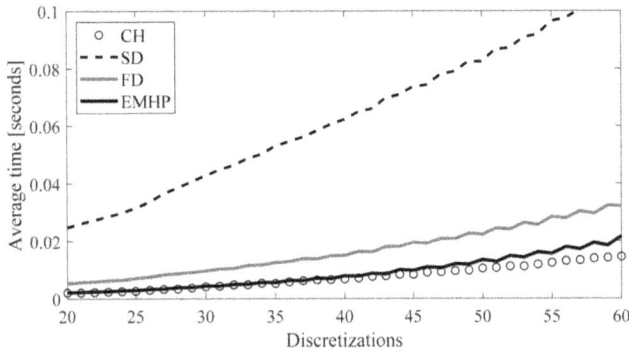

Figure 7. Computation time for each combination of parameters (n, a_p) of all studied methods.

4. Discussion

Chatter is an old enemy to manufacturers but still strongly affects machining times and workpiece quality. Together with mathematical models, this work proposed a numerical method to reproduce many effects arising in practice, which need to be introduced for an accurate representation of the system's stability. These parameters are common ones, such as the geometry of the end-mill (tool diameter, flutes, helix angle, etc.), but also many others, such as (1) runout, (2) cutting speed, and (3) helix angle.

Particularly in the machining of the aluminum alloy family, the authors found that it is mandatory to incorporate the effects of cutting speed in the calculation of the stability lobes, as these may differ greatly depending on the constant or variable cutting coefficients.

The helix angle did not have an important influence in this chart as the helix pitch is 43.5 mm, higher than the 10 mm limit plotted. However, the runout effect is significant not only in the chatter frequencies but also over the modulus of the multipliers. Notice that these values have been increased towards the stability boundaries, and a new region, or flip lobe (when the multipliers leave the unit circle for the real negative axis) appears in the middle of the Hopf bifurcation at around 3250 rpm. As runouts always appear in the workshop, this proposed model can be easily incorporated into any cutting force model because it is based on an experimental measurement.

The cutting speed has a more predictable influence, since the cutting coefficients decrease while the cutting speed increases, and stability boundaries grow as the spindle speed increases. For instance, at 7500 rpm, the axial depth of the cut limit was predicted to be 5.6 mm using a typical analysis, but increased to 7.8 mm for the improved analysis.

The boundary limits are especially sensitive in the stable zones between consecutive lobes, which are often used to improve chip removal rates. For instance, when the spindle speed was 2500 rpm in Figure 5, the prediction here failed due to the severe transition between stable and unstable behavior (see slope of the diagram).

If the prediction model is to be put to good use, a convenient characterization should be done. This non-trivial aspect was not studied by the scientific literature. Models usually assume cutting coefficients to be constant by characterizing and validating the tool-material pair at the same cutting speeds. However, this is not realistic because the most interesting zones between lobes will vary depending on the system's dynamics. The resulting change in the cutting coefficient due to cutting speed cannot be neglected when milling aluminum alloys.

The EMHP method was compared with other similar and competitive methods in the literature, such as the Chebyshev collocation, semi-discretization, and full-discretization methods. A single degree of freedom model was developed and validated with the mentioned updates. When considering computation time and convergence rate, the EMHP and Chebyshev collocation methods proved to be the most efficient when $N < 60$. In terms of overall accuracy, all four methods resulted in the same stability lobes when $N > 30$, but this may not necessarily be the case for other milling conditions.

5. Conclusions

In this work, we studied how we could realistically deal with stability problems when machining aluminum alloys at different cutting speeds. A 1 degree-of-freedom model was solved by using the homotopic perturbation technique. In this case, the proposed cutting force model considered various cutting coefficients, including runout. Stability lobes with varying cutting coefficients and spindle speeds were calculated and experimentally verified for the first time. The developed method proved to be an efficient technique with respect to other state-of-the-art methods. The main contributions can be summarized as:

- We observed a significant variation in the cutting forces depending on cutting speeds, and a model was thus proposed where the cutting-force coefficients varied depending on the cutting speed. Using an exponential cutting speed model, we described how the cutting-force coefficient decreased as a function of the increase in cutting speed.
- It was experimentally demonstrated that inclusion of the effects of the helix angle, runout, and characterization dependent on the cutting speed allowed for much more precise stability boundaries. Furthermore, typical stability lobes with constant cutting coefficients were found to only be valid for a narrow spindle speed range, meaning that their applicability is not valid in real practice.
- The convergence of the EMHP was compared with other efficient methods, such as the semi-discretization and full-discretization methods. When considering numeric convergence and computation times, the EMHP and Chebyshev proved to be the most efficient methods.

Author Contributions: Conceptualization, methodology, software and writing, D.O. and G.U.; Resources and supervision, A.E.-Z. and L.N.L.d.L.

Funding: The authors wish to acknowledge the financial support received from HAZITEK program, from the Department of Economic Development and Infrastructures of the Basque Government and from FEDER funds. Additional support was provided by the Tecnológico de Monterrey, through the Research Group in Nanomaterials and Devices Design.

Conflicts of Interest: The authors declare no conflict of interest. The founding sponsors had no role in the design of the study; in the collection, analyses, or interpretation of data; in the writing of the manuscript, and in the decision to publish the results.

References

1. Smith, S.; Tlusty, J. Efficient Simulation Programs for Chatter in Milling. *CIRP Ann.* **1993**, *42*, 463–466. [CrossRef]
2. Altintas, Y.; Budak, E. Analytical Prediction of Stability Lobes in Milling. *CIRP Ann.* **1995**, *44*, 357–362. [CrossRef]
3. Davies, M.A.; Pratt, J.R.; Dutterer, B.; Burns, T.J. Stability prediction for low radial immersion milling. *J. Manuf. Sci. Eng.* **2002**, *124*, 217–225. [CrossRef]
4. Merdol, S.D.; Altintas, Y. Multifrequency solution of chatter stability for low immersion milling. *J. Manuf. Sci. Eng.* **2004**, *126*, 459–466. [CrossRef]
5. Budak, E.; Altintas, Y. Analytical prediction of chatter stability in milling—Part 1: General formulation. *J. Dyn. Syst. Meas. Control* **1998**, *120*, 22–30. [CrossRef]
6. Bayly, P.V.; Halley, J.E.; Mann, B.P.; Davies, M.A. Stability of interrupted cutting by temporal finite element analysis. *J. Manuf. Sci. Eng.* **2003**, *125*, 220–225. [CrossRef]
7. Insperger, T.; Stepan, G. Semi-discretization method for delayed systems. *Int. J. Numer. Methods Eng.* **2002**, *55*, 503–518. [CrossRef]
8. Insperger, T.; Stepan, G. Updated semi-discretization method for periodic delay-differential equations with discrete delay. *Int. J. Numer. Methods Eng.* **2004**, *61*, 117–141. [CrossRef]

9. Insperger, T.; Stepan, G.; Turi, J. Comparison of zeroth-and first-order semi-discretizations, for the delayed Mathieu equation. In Proceedings of the 43rd IEEE Conference on Decision and Control, Nassau, Bahamas, 14–17 December 2004; Volume 3, pp. 2625–2629.
10. Insperger, T.; Stepan, G.; Turi, J. On the higher-order semi-discretizations for periodic delayed systems. *J. Sound Vib.* **2008**, *313*, 334–341. [CrossRef]
11. Butcher, E.A.; Nindujarla, P.; Bueler, E. Stability of up-and down-milling using Chebyshev collocation method. In Proceedings of the ASME International Design Engineering Technical Conferences and Computers and Information in Engineering Conference, Long Beach, CA, USA, 24–28 September 2005; Volume 6, pp. 841–850.
12. Urbikain, G.; López de Lacalle, L.N.; Fernández, A. Regenerative vibration avoidance due to tool tangential dynamics in interrupted turning operations. *J. Sound Vib.* **2014**, *333*, 3996–4006. [CrossRef]
13. Urbikain, G.; Olvera, D.; López de Lacalle, L.N.; Elías-Zúñiga, A. Spindle speed variation technique in turning operations: Modeling and real implementation. *J. Sound Vib.* **2016**, *383*, 384–396. [CrossRef]
14. Urbikain, G.; Olvera, D.; López de Lacalle, L.N.; Elías-Zúñiga, A. Stability and vibrational behaviour in turning processes with low rotational speeds. *Int. J. Adv. Manuf. Technol.* **2015**, *80*, 871–885. [CrossRef]
15. Ding, Y.; Zhu, L.; Zhang, X.; Ding, H. A full-discretization method for prediction of milling stability. *Int. J. Mach. Tools Manuf.* **2010**, *50*, 502–509. [CrossRef]
16. Insperger, T. Full-discretization and semi-discretization for milling stability prediction: Some comments. *Int. J. Mach. Tools Manuf.* **2010**, *50*, 658–662. [CrossRef]
17. Ding, Y.; Zhu, L.; Zhang, X.; Ding, H. Second-order full-discretization method for milling stability prediction. *Int. J. Mach. Tools Manuf.* **2010**, *50*, 926–932. [CrossRef]
18. Compeán, F.I.; Olvera, D.; Campa, F.J.; López de Lacalle, L.N.; Elías-Zúñiga, A.; Rodríguez, C.A. Characterization and stability analysis of a multivariable milling tool by the enhanced multistage homotopy perturbation method. *Int. J. Mach. Tools Manuf.* **2012**, *57*, 27–33. [CrossRef]
19. Schmitz, T.L.; Couey, J.; Marsh, E.; Mauntler, N.; Hughes, D. Runout effects in milling: Surface finish, surface location error, and stability. *Int. J. Mach. Tools Manuf.* **2007**, *47*, 841–851. [CrossRef]
20. Engin, S.; Altintas, Y. Mechanics and dynamics of general milling cutters. Part I: Helical end mills. *Int. J. Mach. Tools Manuf.* **2001**, *41*, 2195–2212. [CrossRef]
21. Mann, B.P.; Insperger, T.; Bayly, P.V.; Stepan, G. Stability of up-milling and down-milling, part 2: Experimental verification. *Int. J. Mach. Tools Manuf.* **2003**, *43*, 35–40. [CrossRef]
22. Astakhov, V.P. *Tribology of Metal Cutting*, 1st ed.; Elsevier Science: New York, NY, USA, 2006; Volume 52.
23. Kumme, R.; Mack, O.; Bill, B.; Gossweiler, C.; Haab, H.R. Dynamic Properties and Investigations of Piezoelectric Force Measuring Devices. *VDI BERICHTE* **2002**, *1685*, 161–172.

applied
sciences

MDPI

Article

Effect of Droplet Impingement on the Weld Profile and Grain Morphology in the Welding of Aluminum Alloys

Zhanhui Zhang, Jiaxiang Xue *, Li Jin and Wei Wu

School of Mechanical and Automotive Engineering, South China University of Technology,
Guangzhou 510641, China; mezhangzh@mail.scut.edu.cn (Z.Z.); jinli8756@163.com (L.J.);
201710100398@mail.scut.edu.cn (W.W.)
* Correspondence: mejiaxue@scut.edu.cn; Tel.: +86-020-2223-6360

Received: 3 July 2018; Accepted: 19 July 2018; Published: 23 July 2018

Featured Application: This work may be used to improve the weld bead quality affected by the inclination of welding torch.

Abstract: To achieve a better understanding of the effect of droplet impingement on the weld profile and grain morphology, welding with vertical and inclined torches in the double pulsed-gas metal arc welding of aluminum alloy were compared. When using vertical welding, the grains along the wall of the finger-like penetration (FLP) were refined by a more violent flow driven by droplet impingement running in the confined space created by FLP. When using inclined welding, the sharp inflection point disappeared and the curved columnar grains emerged on the non-impact action side, which was attributed to the gradually weakened impingement at that location. Moreover, when the penetration became shallower due to a low mean current, the droplets impinged alternately along split trajectories, causing significant changes in the grain morphology, such as creating grains which were sharply shortened by the direct impact of droplet impingement at impact point. The change of trajectory was ascribed to the variation of the width/depth ratio of FLP, which changed the magnitude of the contradiction between the room required by the fluid flow driven by droplet impingement and the space supplied for that by FLP.

Keywords: welding; aluminum alloy; refinement; droplet impingement; weld pool behavior

---

1. Introduction

The, 6XXX series of aluminum alloys are widely used in different fields because of their high strength-to-weight ratio [1,2]. These components are often fabricated into industrial structures by fusion welding processes [2]. Gas metal arc welding (GMAW) is an important fusion welding method of joining aluminum alloys [3]. In GMAW, the weld profile and microstructure are strongly affected by the molten pool behavior [4], which in turn is strongly affected by various driving forces, such as the arc pressure, plasma shear force, Marangoni force and electromagnetic force (EMF) [4,5]. However, in GMAW, the molten pool behavior is more complex because it involves the impingement of molten droplets into the weld pool, relative to gas tungsten arc welding which has no droplet impingement [4,6]. Hu and Tsai [6] found that a detached droplet is accelerated by the plasma arc during its flight toward the workpiece. When a droplet was transferred into the molten pool, the flow pattern was mainly determined by the droplet impact [4,7]. Cao et al. [8] simulated two different metal droplet velocities and found that the weld penetration was primarily determined by the metal droplet impact force. Moreover, by developing a unified model considering both the heat and mass transfer in the electrode, arc plasma and molten pool, Fan and Kovacevic [9] found that the penetration of the weld pool was mainly determined by the impingement of droplets. After droplet impingement, a "crater" is created in the weld pool [10], and the finger-like penetration (FLP) geometry of the weld bead is formed [8,11]. As shown in the lower right corner of Figure 1, the weld profile with the FLP in the transverse section was characterized by a shallow and wide penetration in the upper region, labeled zone A in Figure 1, and a deep and narrow penetration in the lower weld [12], labeled zone B in Figure 1. Zone B is known as FLP.

Although the geometric characteristics of the FLP are well recognized, the grain morphology of this type of weld bead has not been fully investigated. Previous research has mainly focused on various methods for the refinement of grains, such as ultrasonic stirring [13], pulsed ultrasonic stirring [14], electromagnetic stirring [15], and arc oscillation [16], and grains in the central part of a weld bead which were often set as the focus and evaluated indicator for the validity of a method. This may be because grains along a weld wall have been widely recognized as having a columnar morphology [17,18]. However, a weld bead with FLP characteristics displays a quite different result in this study. So far, the difference in grain morphology between the walls of the FLP and the walls of the shallow region has not been reported, nor has the mechanism causing the variation.

As mentioned above, droplet impingement plays a significant role in determining the flow pattern, temperature distribution and resultant weld profile and grain morphology. In this study, we focus on the weld profile and grain morphology of aluminum alloys based on droplet impingement with the welding torch held perpendicular to the base metal. Subsequently, to investigate the effect of the impingement direction, the welding torch is held at a 10° inclination in the transverse section. Finally, to further understand how droplet impingement is affected by penetration depth, the double pulsed-gas metal arc welding (DP-GMAW) method, which can produce both deep and shallow penetration in a single travel, was employed.

2. Materials and Methods

Bead-on-plate welding experiments were conducted on aluminum alloy AA6061 base metal (250 × 60 × 3 mm³) using ER4043 filler wire. The chemical compositions of alloys AA6061 and ER4043, are provided in Table 1 [1]. The DP-GMAW process parameters are given in Table 2. Pure argon was coaxially supplied as a shielding gas at a flow rate of 18 L/min. After welding, the specimens were cut along the deepest and shallowest transverse cross sections of the welds, as schematically shown by the center line in the lower part of Figure 1. The samples were obtained from 3 cases, as shown in Figure 1, case 1: the sample with deepest penetration and welding torch perpendicular to the base metal; case 2: the sample with deepest penetration and a 10° inclination of the welding torch to the base metal; and case 3: the sample with shallowest penetration and a 10° inclination of the welding torch to the base metal. The samples were ground and polished with colloidal silica and were subsequently etched for metallographic analysis using the standard Keller agent for 20 s.

DP-GMAW was employed to obtain different depth penetrations under the same condition. DP-GMAW was characterized by its special current waveform with two sets of pulsing current: the first phase (namely the thermal peak), with a high mean current, produced deep penetration, and the second phase (namely the thermal base), with a low mean current, produced shallow penetration. The current waveform and temporal variations of the longitudinal sections of the welds produced by DP-GMAW, are schematically illustrated on the left-hand side of Figure 1, from which it can be seen clearly that the penetration increased when welding with a high mean current in the first phase, and decreased when welding with a low mean current in the second phase.

Table 1. Chemical compositions of AA6061 and ER4043 (wt%) [1].

Material	Mg	Si	Fe	Cu	Mn	Cr	Al
AA6061	1.02	0.75	0.45	0.25	0.06	0.05	Bal.
ER4043	0.05	5.60	0.80	0.30	0.05	–	Bal.

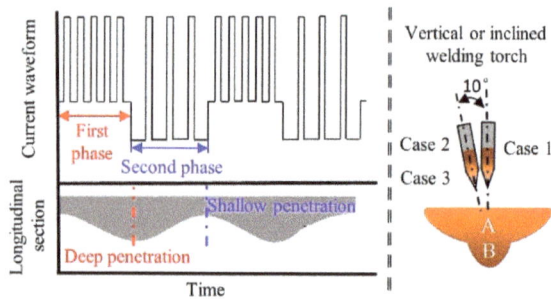

Figure 1. Schematic diagram of the current waveform and longitudinal sections of the double pulsed-gas metal arc welding (DP-GMAW) weld pool and the inclination of the welding torch.

Table 2. Welding process parameters for DP-GMAW.

Process Parameters	Value
Mean voltage (V)	21.7
Mean current (A)	96
Wire diameter (mm)	1.2
Welding speed (mm/s)	59
Wire feeding rate (mm/s)	2.8
First phase current (A)	114
Second phase current (A)	78
First phase time (ms)	180
Second phase time (ms)	180

3. Results and Discussion

3.1. Welding Torch Perpendicular to the Base Metal

When the welding torch was perpendicular to the top surface of the base metal in the transverse section (case 1), FLP was commonly observed in the fusion zone [8,12], as shown in Figure 2a. The formation of the weld profile with FLP was mainly due to the combination of the outward flow of the molten metal at a shallow level, and downward flow at a deep level. The outward flow was responsible for the wide but shallow penetration, and the downward flow was responsible for the deep and narrow penetration. Under the combined driving forces, including a drag force induced by the difference in tangential velocity between the plasma and the molten pool, and a Marangoni force caused by surface tension and its spatial gradient [5,19], the molten metal flowed outward. Next, more of the solid base metal at the liquid/solid interface was molten, thereby increasing the width of the weld pool. The heat input occurred over a large area, which could not melt deeper because of the lack of a powerful downward flow. Therefore, a wide but shallow concave region formed.

The remarkable surface depression caused by arc pressure following a Gaussian distribution thinned the fluid layer between the arc and solid base metal [20], thus decreasing the buffering of the weld puddle from the droplet impact [4,21]. The acceleration of the droplets due to the plasma drag force [6], combined with the weakened buffering effect of the weld puddle, made it easier for droplets to reach the bottom boundary of the weld pool, even with substantial redundant momentum. Moreover, because of the heat input before droplet impingement, the solid bottom was heat-softened [22]. Thus, when an accelerated and superheated droplet impinges on a thin fluid layer, it flows downward and crashes against the solid but already-softened bottom; creating a crater. Note that the heat and momentum carried by droplets weakened and the size of the droplets decreased as they flowed downward, and after crashing, the droplets scattered into pieces to flow to both sides [23], as represented by the green dashed arrows in Figure 2b. Then, the crater became increasingly deep under the successive impingement of droplets, as shown by the black dashed line in Figure 2b. Finally, a deep penetration formed. Meanwhile, because of the significant difference in velocity between the droplet and the neighboring liquid in the molten pool and the reduced viscosity caused by increased temperature, only a small fraction of the fluid near the droplets was heated and accelerated, and this fluid could only melt a narrower area than at the shallow level. Therefore, a distinct separating landmark, namely a sharp inflection point (SIP), formed between the shallow and the FLP levels.

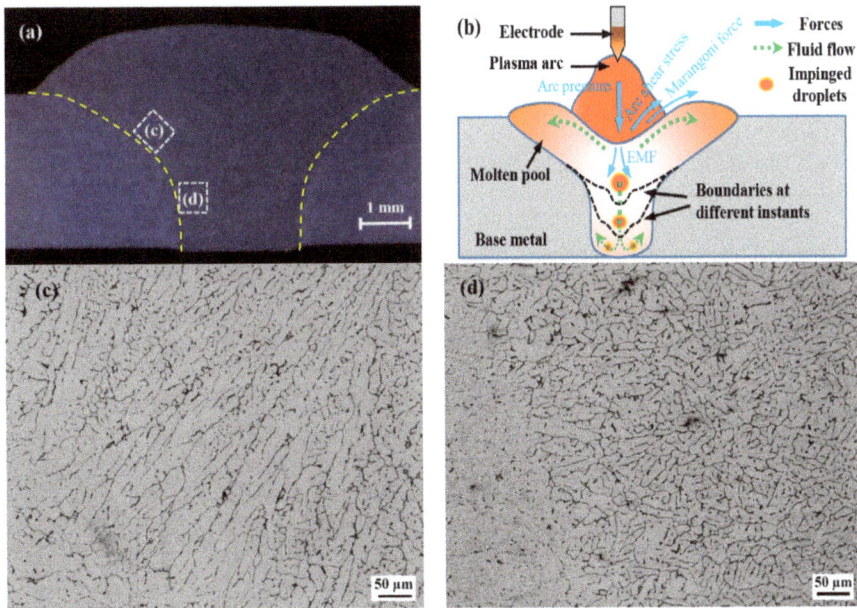

Figure 2. Welding torch perpendicular to the surface of the base metal in the transverse section. (**a**) Molten pool profile; (**b**) schematic diagram of the molten pool behavior; (**c**) columnar grain at the shallow level; (**d**) refined grain at the finger-like penetration (FLP) level.

The grain morphology along the different walls of the molten pool differed considerably, including both columnar grains (Figure 2c) at the shallow level, and refined equiaxed dendrites (Figure 2d) at the FLP level. The difference in grain morphology could be ascribed to the contradiction between the room required by the momentum obtained by the molten metal from droplet impingement and the space which could be provided by the FLP. When the weld pool provides space far beyond that required, the fluid carrying heat flux does not have sufficient momentum to run violently throughout the space; thus, the growth of grains will not be interrupted. In contrast, when the molten metal acquires more momentum to flow but only a small region can be supplied for that flow by the molten pool, then the grain growth will be disturbed and refined by the momentum and heat which is brought by droplet impingement. Because many researchers have provided details of how fluid flow refines grains, as was determined in References [14,16], a detailed description of the process was not provided here. As shown in Figure 2b, at the shallow level, the weld pool was sufficiently wide that the fluid flow which was driven by the combined drag force and Marangoni force was not sufficiently powerful to run violently throughout the shallow level, particularly at the walls. Thus, the grains grew into a columnar morphology without interruption by the fluid flow. Compared to the redundant room for flow at the shallow level, the flow of molten metal was limited to a deep and narrow space. In addition to the limited space, the downward flow was driven by the powerful high-speed droplets [4,8]; thus, the contradiction was intensified significantly, and the more powerful fluid had the ability to run violently throughout a limited FLP. Subsequently, the grain growth was continuously interrupted, causing grains along the walls of the FLP to refine and transit to equiaxed dendrites.

3.2. Welding Torch Inclined in the Transverse Section

The profile of the weld pool was significantly different when the welding torch was set at an angle of 10° from the center line in the transverse section (case 2), with the disappearance of the SIP on the non-action side (Figure 3a). This phenomenon possibly arose because when the welding torch is inclined, heat input becomes asymmetric about the longitudinal central plane, outlined in References [24,25], as shown in Figure 3b. Because of the asymmetric heat input, more base metal melted on the left side of the axial line of the welding wire than on the right side, causing the bottom to be uneven. Thus, when the droplet impacted the uneven bottom of the weld pool, the more molten region (in this case, the left side) had a considerably lower resistance to the impingement because of its liquid state than the solid-state region on the right side [11,26]. Thus, instead of impacting both sides at the same order of magnitude, as in case 1, the droplet impacted harder on the left side due to its weak resistance. The original uneven solid bottom became more uneven because of the additional hard impact and heat input from droplet impingement. Next, an asymmetric crater formed because the left side bottom was considerably lower than the right side bottom; this crater could serve as a confined channel to redirect the droplet impingement [27]. Then, when another droplet impinged, it tended to crash heavily against the solid walls on the impact action side (IAS), thus melting and paring the solid-state wall, whilst flowing downward until finally smashing against the bottom. The dimensions of the FLP were measured with 2.849 mm in width and 1.785 mm in depth, and the width-to-depth aspect ratio was 1.596. Next, the boundaries were expanded downward in the same mode, as depicted by the black dashed lines in Figure 3b. However, there was no crashing against the other side wall during the entire downward flow process because its liquid state facilitated the flowability for droplet impingement. This lack of crashing may have accounted for the remnant SIP on the IAS, and the disappearance of SIP on the non-impact action side (NIAS).

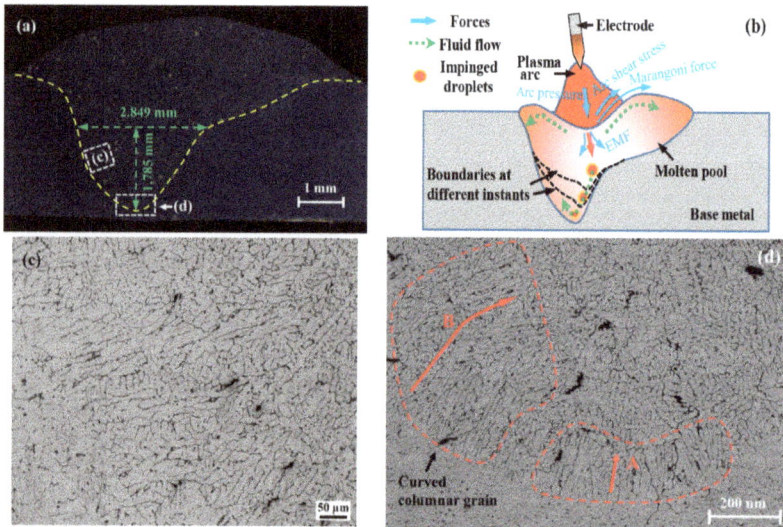

Figure 3. Welding torch inclined from the center line in the transverse section. (**a**) Molten pool profile with the sharp inflection point (SIP) not present on the non-impact action side (NIAS); (**b**) schematic diagram of the molten pool behavior; (**c**) columnar grain on the NIAS in the FLP level; (**d**) shortened columnar grain on the impact action side (IAS) at the shallow level.

Figure 3c,d demonstrates the grain morphology distribution characteristics at different positions of the weld pool walls. The grains along the wall of the IAS in the FLP, where droplet impact occurred, were refined with no discernible difference from the same area in case 1. However, the columnar grains emerged along the walls of the NIAS in the FLP (Figure 3c), whereas refined grains were observed at the same position in case 1. This phenomenon may have occurred because both the momentum and heat obtained from the superheated droplets were consumed most at, and shortly after crashing against the wall of the IAS until they reached the bottom. This interpretation is supported by Figure 3d, which clearly demonstrates the variation trend of columnar grains in the bottom region. The figure shows that the columnar grains with small sizes emerged in the bottom region (zone A in Figure 3d) because of the weak existence of droplet impingement. Afterward, when the fluid flowed to the NIAS region, the droplet-impingement-driven-flow had nearly exhausted of all its momentum and heat, losing its ability to run violently throughout the FLP region, thus providing the grains in this region with the opportunity to grow without interruption into a columnar morphology with greater size in the NIAS region (zone B in Figure 3d). The columnar grains in the NIAS region grew in a curved pattern consisting of a multi-segment line, rather than along a single straight line from the fusion boundary toward the center of the molten pool. DebRoy and his coworkers [23] claimed that the local solidification direction of a columnar grain changing with the movement of the molten pool was responsible for this phenomenon. In contrast to their experiment with gas tungsten arc welding, there was droplet impingement involved in this study, which played a critical role in determining the direction in which the temperature gradient had the maximum value. Hence, in this study, the droplet impingement was likely responsible for the emergence of curved grains due to its role in changing the temperature gradient and thus the local solidification direction.

3.3. Shallow Penetration with an Inclined Torch

When the welding current entered into the small current phase with an inclined welding torch (case 3), the weld penetration depth decreased from 1.785 mm to 1.259 mm, whilst the width increased slightly from 2.849 mm to 2.871 mm compared with case 2, as shown by the dimensions of the FLP in Figure 3a for case 2 and Figure 4a for case 3. This significant decrease in penetration could be ascribed to the change in flow pattern in the molten pool, as schematically illustrated by Figure 4b. The figure shows that the trajectory of flow driven by droplet impingement was split into two branches at the SIP when the penetration depth decreased: (i) a branch flowing downward to the bottom, as depicted by the red curved arrows which represented the signs of the trajectories created by repeated droplet impingement (Figure 4c), and (ii) another branch flowing upward along the wall of the shallow level, as indicated by the gradually enlarged grain size with increased distance from the SIP (Figure 4d), which could also be interpreted as the outline of many droplet impingements. This change in flow pattern may have resulted from the constantly varying contradiction between the space provided for liquid flow and the space required by the momentum of flow propelled by droplet impingement. When the droplet impinged into the weld pool, it impacted and peeled the walls and then crashed against the bottom, as in case 2. However, because of the reduced penetration depth, the room supplied was far less than required, instantly intensifying the contradiction dramatically. As a result, the shrunken FLP region could no longer hold the violent flow, which caused molten metal to overflow the FLP region [6,24], as schematically illustrated by the dashed circle arrow in Figure 4b. The overflowing effect could be proven by the width-to-depth aspect ratio [7]. In this study, the FLP aspect ratio increased from 1.596 to 2.281, an increase of 42.9 percent, due to the width being enlarged slightly, whereas the penetration was dramatically reduced by 41.7 percent compared to case 2, as shown in Figure 4e. The overflowing fluid then further exerted its influence on the trajectory of the impinging droplets by pushing them aside to the upward direction (Figure 4b). In the shallow level, the fluid flowed upward along the wall driven by electromagnetic forces (EMF) [25]; this flow may have acted as an inducing flow for droplet impingement. Hence, once pushed aside, the droplets were more involved in, and ultimately converged with the upward flow driven by EMF, as indicated by upward black dashed arrows.

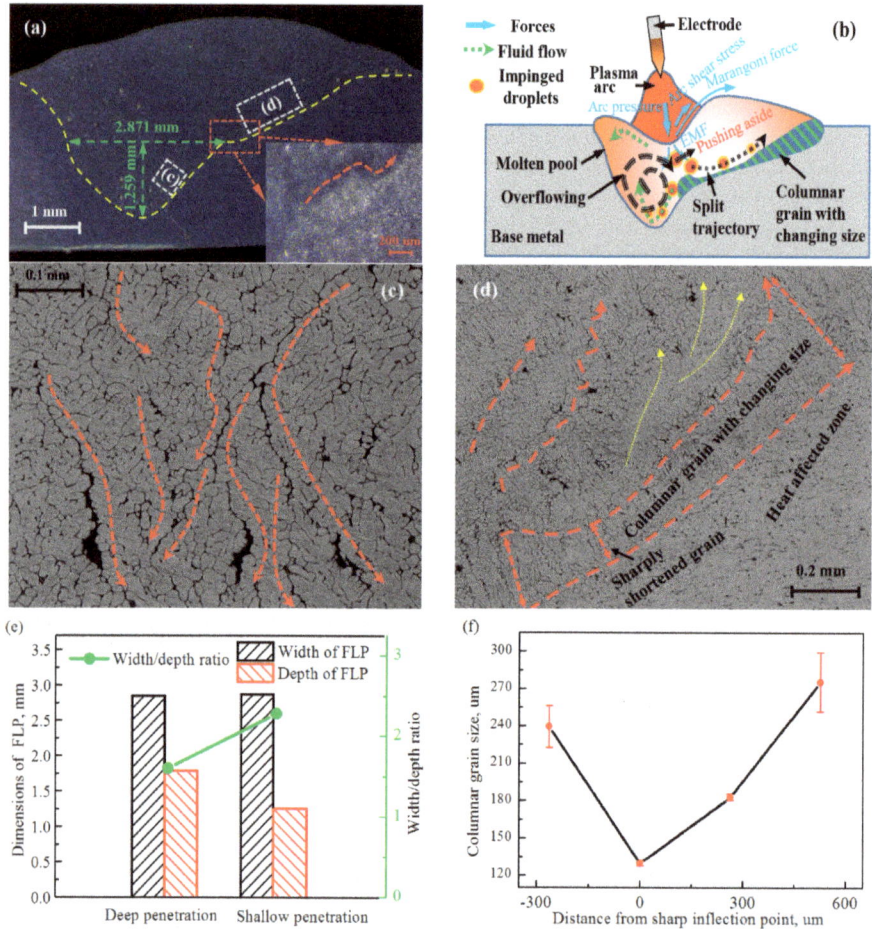

Figure 4. Welding torch inclined from the center line in the transverse section. (**a**) Molten pool profile; (**b**) schematic diagram of the molten pool behavior; (**c**) signs of downward flow; (**d**) signs of upward flow; (**e**) width, depth and width/depth ratio of the FLP; (**f**) columnar grain size along the upward wall.

The upward flow shaped the grain morphology along the wall into three stages (Figure 4d): large size before the impact point (IP), reduced size at the IP and in a near neighboring region, and large size again far after the IP. The variation in the grain size with increased distance from the SIP, as illustrated by Figure 4f, may have resulted from the corresponding change in flow magnitude [3]. Different regions suffered different order-of-magnitude impacts and heat inputs from the droplets. The area before the IP suffered slightly and then the grain grew up, whereas the size of the grains at the IP decreased sharply to approximately half of that in the area before the IP, as shown by the local magnified image in Figure 4a and the optical micrograph in (Figure 4f,) which may have been because this area was impacted directly and heavily by droplet impingement, which carried abundant heat and momentum. From another perspective, the shortened columnar grains in the IP region along the upward wall (Figure 4d) provided evidence that droplet impingement had the ability to refine grains directly through impact. Because of the gradually attenuated impact and heat with increasing distance from the IP, the grain size increased again. In fact, the outline of the changing grain size was the trajectory of droplet impingement, as indicated by the other red dashed arrows in Figure 4d. However, the outline was not shaped into the final curve by one single droplet impingement; each time a droplet impinged, the outline was prolonged slightly, as indicated by the yellow curved arrows in Figure 4d. After many droplet impingements, the outline accumulated to the final curve.

When the trajectory of droplet impingement shifted upward, the violent flow in the FLP level gradually diminished because of the lack of sustained bulk momentum input [24], which in turn lowered the space requirement, and reduced the contradiction between the generated and required space. As a result, space with the same size in the FLP level regained the ability to accommodate the reduced magnitude of liquid flow, ending the overflow. This lack of overflow diminished the effect of pushing aside, and thus, the droplet suffering non-pushed effects impinged back into the FLP region again. Then, the already-weakened violent flow in the FLP level was strengthened again, as was the subsequent effect of pushing aside. Therefore, the trajectory of droplet impingement was pushed upward once more. As a result, the cyclic alternative shift between the downward and upward directions was triggered. When droplets were flowing downward, the grains in the upper molten pool continued to grow, which may have accounted for the distance between the curved arrows in Figure 4d.

The grains at the SIP between the separate trajectories survived to grow into a columnar morphology with a considerably greater size than in the neighboring region, as presented by the local magnified image of the region, indicated by the red rectangle in Figure 4a. This enlargement may have been due to the sensitivity of the droplet trajectory to the effect of pushing aside. When the effect of pushing aside was strengthened slightly by the enhanced overflow, the trajectory was shifted significantly in the upward direction. Conversely, the slightly weakened effect pulled droplet impingement considerably downward. Therefore, the region between the trajectories remained unaffected, and avoided a severe impact and heat input from powerful droplet impingement. Hence, the larger columnar grains survived this process.

4. Conclusions

In summary, our findings revealed the significant role of droplet impingement in determining the weld pool profile and grain morphology in the welding of aluminum alloys. The main conclusions were as follows:

1. The grains along the wall of the FLP region were more refined than those in the shallow region because of the more violent flow driven by droplet impingement running in a confined space created by the FLP. In summary, grain morphology is determined by the magnitude of contradiction between the room required by fluid flow and the space supplied by the weld pool.
2. When the torch is inclined, the SIP disappears and the curved columnar grains emerge along the NIAS wall because of the gradually weakened impingement at that location. Then, when the penetration becomes shallow, the trajectory of droplet impingement is split into two branches under the combined effect of pushing aside and inducing flow which is driven by EMF.
3. Grains along the wall of the upward branch undergo three stages due to the change in momentum and heat carried by droplets with increasing distance. However, grains before the IP survive to grow into a columnar morphology with a markedly larger size than in the neighboring region because the columnar grain region may not be impacted by droplets, due to the sensitivity of the impingement trajectory to the effect of pushing aside.
4. Droplet impingement can refine grains directly through impact, as evidenced by the sharply shortened columnar grain at the impact point in case 3.

Author Contributions: Z.Z. and J.X. conceived and designed the experiments, and wrote the paper; L.J. and W.W. performed the experiments; and all the authors analyzed and discussed the data.

Funding: This work was supported by the High-level Leading Talent Introduction Program of GDAS (No. 2016GDASRC-0106), the National Natural Science Foundation of Guangdong Province (No. 2016A0303117), and Natural Science Foundation of Fujian (No. 2018J01503).

Conflicts of Interest: The authors declare no conflict of interest.

References

1. Bardel, D.; Fontaine, M.; Chaise, T.; Perez, M.; Nelias, D.; Bourlier, F.; Garnier, J. Integrated modelling of a 6061-T6 weld joint: From microstructure to mechanical properties. *Acta Mater.* **2016**, *117*, 81–90. [CrossRef]
2. Ambriz, R.R.; Mesmacque, G.; Ruiz, A.; Amrouche, A.; López, V.H. Effect of the welding profile generated by the modified indirect electric arc technique on the fatigue behavior of 6061-T6 aluminum alloy. *Mater. Sci. Eng. A* **2010**, *527*, 2057–2064. [CrossRef]
3. Wang, L.L.; Wei, H.L.; Xue, J.X.; DebRoy, T. A pathway to microstructural refinement through double pulsed gas metal arc welding. *Scr. Mater.* **2017**, *134*, 61–65. [CrossRef]
4. Rao, Z.H.; Zhou, J.; Liao, S.M.; Tsai, H.L. Three-dimensional modeling of transport phenomena and their effect on the formation of ripples in gas metal arc welding. *J. Appl. Phys.* **2010**, *107*, 054905. [CrossRef]
5. Meng, X.; Qin, G.; Zou, Z. Sensitivity of driving forces on molten pool behavior and defect formation in high-speed gas tungsten arc welding. *Int. J. Heat Mass Transf.* **2017**, *107*, 1119–1128. [CrossRef]
6. Hu, J.; Tsai, H.L. Heat and mass transfer in gas metal arc welding. Part II: The metal. *Int. J. Heat Mass Transf.* **2007**, *50*, 808–820. [CrossRef]
7. Fan, H.G.; Kovacevic, R. A unified model of transport phenomena in gas metal arc welding including electrode, arc plasma and molten pool. *J. Phys. D Appl. Phys.* **2004**, *37*, 2531–2544. [CrossRef]
8. Davies, M.H.; Wahab, M.; Painter, M.J. An Investigation of the Interaction of a Molten Droplet with a Liquid Weld Pool Surface: A Computational and Experimental Approach. *Weld. J.* **2000**, *79*, 18s–23s.
9. Cao, Z.; Yang, Z.; Chen, X.L. Three-dimensional simulation of transient GMA weld pool with free surface. *Weld. J.* **2004**, *83*, 169s–176s.
10. Wang, Y.; Tsai, H.L. Impingement of filler droplets and weld pool dynamics during gas metal arc welding process. *Int. J. Heat Mass Transf.* **2001**, *44*, 2067–2080. [CrossRef]
11. Kim, C.H.; Zhang, W.; DebRoy, T. Modeling of temperature field and solidified surface profile during gas–metal arc fillet welding. *J. Appl. Phys.* **2003**, *94*, 2667–2679. [CrossRef]
12. Cheon, J.; Kiran, D.V.; Na, S.J. CFD based visualization of the finger shaped evolution in the gas metal arc welding process. *Int. J. Heat Mass Transf.* **2016**, *97*, 1–14. [CrossRef]
13. Yuan, T.; Kou, S.; Luo, Z. Grain refining by ultrasonic stirring of the weld pool. *Acta Mater.* **2016**, *106*, 144–154. [CrossRef]
14. Wang, J.; Sun, Q.; Liu, J.; Wang, B.; Feng, J. Effect of pulsed ultrasonic on arc acoustic binding in pulsed ultrasonic wave-assisted pulsed gas tungsten arc welding. *Sci. Technol. Weld. Join.* **2016**, *22*, 465–471. [CrossRef]
15. Li, Y.; Zhang, Y.; Bi, J.; Luo, Z. Impact of electromagnetic stirring upon weld quality of Al/Ti dissimilar materials resistance spot welding. *Mater. Des.* **2015**, *83*, 577–586. [CrossRef]
16. Yuan, T.; Luo, Z.; Kou, S. Mechanism of grain refining in AZ91 Mg welds by arc oscillation. *Sci. Technol. Weld. Join.* **2016**, *22*, 97–103. [CrossRef]
17. Kumar, R.; Dilthey, U.; Dwivedi, D.K.; Ghosh, P.K. Thin sheet welding of Al 6082 alloy by AC pulse-GMA and AC wave pulse-GMA welding. *Mater. Des.* **2009**, *30*, 306–313. [CrossRef]
18. Wei, H.L.; Elmer, J.W.; DebRoy, T. Origin of grain orientation during solidification of an aluminum alloy. *Acta Mater.* **2016**, *115*, 123–131. [CrossRef]
19. Fan, H.G.; Kovacevic, R. Droplet Formation, Detachment, and Impingement on the Molten Pool in Gas Metal Arc Welding. *Metall. Mater. Trans. B* **1999**, *30*, 791–801. [CrossRef]
20. Tanaka, M.; Terasaki, H.; Ushio, M.; Lowke, J.J. A unified numerical modeling of stationary tungsten-inert-gas welding process. *Metall. Mater. Trans. A* **2002**, *33*, 2043–2052. [CrossRef]
21. Wang, L.L.; Lu, F.G.; Cui, H.C.; Tang, X.H. Investigation of molten pool oscillation during GMAW-P process based on a 3D model. *J. Phys. D Appl. Phys.* **2014**, *47*, 465204. [CrossRef]

22. Cai, X.Y.; Lin, S.B.; Fan, C.L.; Yang, C.L.; Zhang, W.; Wang, Y.W. Molten pool behaviour and weld forming mechanism of tandem narrow gap vertical GMAW. *Sci. Technol. Weld. Join.* **2016**, *21*, 124–130. [CrossRef]
23. Guo, H.; Hu, J.; Tsai, H.L. Formation of weld crater in GMAW of aluminum alloys. *Int. J. Heat Mass Transf.* **2009**, *52*, 5533–5546. [CrossRef]
24. Cho, D.W.; Song, W.H.; Cho, M.H.; Na, S.J. Analysis of submerged arc welding process by three-dimensional computational fluid dynamics simulations. *J. Mater. Process. Technol.* **2013**, *213*, 2278–2291. [CrossRef]
25. Cho, D.W.; Na, S.J. Molten pool behaviors for second pass V-groove GMAW. *Int. J. Heat Mass Transf.* **2015**, *88*, 945–956. [CrossRef]
26. Liu, J.W.; Rao, Z.H.; Liao, S.M.; Tsai, H.L. Numerical investigation of weld pool behaviors and ripple formation for a moving GTA welding under pulsed currents. *Int. J. Heat Mass Transf.* **2015**, *91*, 990–1000. [CrossRef]
27. Wang, L.; Wu, C.S.; Gao, J.Q. Suppression of humping bead in high speed GMAW with external magnetic field. *Sci. Technol. Weld. Join.* **2016**, *21*, 131–139. [CrossRef]

applied
sciences

MDPI

Article

A New Technique for Batch Production of Tubular Anodic Aluminum Oxide Films for Filtering Applications

Chien Wan Hun [1], Yu-Jia Chiu [2], Zhiping Luo [3], Chien Chon Chen [4],* and Shih Hsun Chen [5,6],*

[1] Department of Mechanical Engineering, National United University, Miaoli 36003, Taiwan;
     monger@nuu.edu.tw
[2] Hydrotech Research Institute, National Taiwan University, Taipei 10617, Taiwan; yujiachiu@ntu.edu.tw
[3] Department of Chemistry and Physics, Fayetteville State University, Fayetteville, NC 28301, USA;
     zluo@uncfsu.edu
[4] Department of Energy Engineering, National United University, Miaoli 36003, Taiwan
[5] Department of Mechanical Engineering, National Taiwan University of Science and Technology,
     Taipei 10607, Taiwan
[6] Department of Materials Science and Engineering, Northwestern University, Evanston, IL 60208, USA
*   Correspondence: chentexas@gmail.com (C.C.C.); shchen@mail.ntust.edu.tw (S.H.C.)

Received: 1 May 2018; Accepted: 26 June 2018; Published: 28 June 2018

Abstract: With larger surface areas and nanochannels for mass delivery and gas diffusion, three-dimensional tubular anodic aluminum oxide (AAO) films have practical advantages over two-dimensional AAO films for medical and energy applications. In this research, we have developed a process for batch production of tubular AAO films using a 6061 Al tube. The tubular AAO films have open nano-channels on both sides, with average pore dimensions of about 60 nm and pore densities of about 10^8 to 10^9 pore/cm^2. It was found that the porous AAO material with nano-channel structure exhibited dialysis behavior, allowing for liquid/gas exudation through diffusion between the inner and outer surfaces of the tubular AAO films. Ar gas bearing test and aeration test were conducted to find the pressure bearing capacity of tubular AAO films. It was demonstrated that the AAO film with a thickness of 100 μm can resist an argon pressure up to 8 atm; however, 30 μm AAO film can only withstand 3 atm of Ar gas. The tubular AAO films with exudation characteristics have the potential for applications in advanced technologies, such as liquid or gas filters, drug delivery, and energy applications.

Keywords: tubular anodic aluminum oxide (AAO); porous material; nanochannels; filtering

1. Introduction

High-quality AAO (anodic aluminum oxide) films provide ordered straight channels, with a diameter of 10–500 nm, pore density of 10^7–10^{11} pore/cm^2, and thickness of 1–300 μm [1,2]. With large surface areas, high mechanical strength, and flexibility, AAO can be used in medical or energy applications, such as drug delivery and detection [3,4]. The large AAO surfaces can be utilized to absorb the bio-indicators or drugs, and the releasing behavior can also be controlled based on the heat sensitivity. AAO has also found applications in energy conversion between carbon dioxide (CO_2) and methane (CH_4) [5,6]. By loading photocatalyst particles on the AAO surface, such photocatalytic systems can be used to recycle carbon dioxide into organic compounds. Based on the features of larger surface areas and nanochannels for mass delivery and gas diffusion, three-dimensional (3D) structure of AAO films have practical advantages over two-dimensional (2D) AAO films for medical and energy applications.

AAO has a lower melting point than pure alumina because of the inclusions in the porous AAO structure. Spooner [7] presented the following compositions of alumina film anodized using sulfuric acid as electrolyte: Al_2O_3 (78.9 wt. %), $Al_2O_3 \cdot H_2O$ (0.5 wt. %), $Al_2 (SO_4)_3$, and H_2O (0.4 wt. %). According to Akahori's work [8], the melting point of AAO is near 1200 °C, and AAO template retains stable, at around 1000 °C [9], which is lower than that of bulk alumina (2017 °C for $Al_2O_{3(\gamma)}$). The AAO structure maintaining temperature of 1000 °C is stable enough to be a template for CaO-$CaCO_3$ reaction at 894 °C.

In the past, tubular AAO has attracted attention. Several methods were proposed for the fabrication and applications of tubular AAO films. According to Altuntas's report [10], the large area (50 cm^2) free-standing AAO membranes was obtained by sputtering carbon onto AAO surface for conductive AAO biosensor applications, especially tubular AAO for filtering. Belwalkar [11] showed that AAO tubular membranes were fabricated from aluminum alloy tubes in sulfuric and oxalic acid electrolytes, the pore sizes were ranging from 14 to 24 nm, and the wall thicknesses was as high as 76 μm, which increased the mechanical strengths for handing. The pore density can be calculated by determining the number of pores according to the area fraction: $P = A_p/[(\pi/4)D^2]$, where P is the number of pores, A_p is the area fraction of pores, and D is the average pore diameter. Gong [12] presented that AAO membrance was prepared in 0.2 M oxalic acid electrolyte under 25 to 40 V applied for 11 to 18 h; additionally, the control in drug delivery by using nanoporous alumina tube capsules with pores of 25 to 55 nm was demonstrated. Kasi1 [13] further reported that the purity of Al is also a matter of great concern for AAO fabrication. Some applications, such as nano-templates for semiconductor industry, require a regular pore arrangement with a honeycomb structure, which cannot be achieved from low grade Al. Moreover, AAO membrane in tubular form can further satisfy the demand in both diffusive and convective filtration of hemodialysis.

The control in preparing tubular AAO is more challenging than flat AAO film on Al sheet because of film cracking issues. Since the AAO processes are sensitive to the operation conditions, defects may appear in AAO membranes if unsuitable conditions are applied. How to manage the variable conditions is pivotal, including electrolyte temperatures, applied voltages, electrolyte compositions, cooling stirring, and current density distribution. Kasi [14] reported that more cracks were generated in smaller diameter (2 mm) tube than larger one (19 mm). The number of cracks generated on the surface of tubular membrane is directly proportional to the thickness of AAO membrane. The result also showed that a crack free thicker membrane could only be obtained if the tube diameter is large. Itoh [15] also shows the AAO tubes that were synthesized by anodization from interior outwards were three times stronger than those anodized from the exterior surface. Grimes [16] demonstrated the nanoporous filter is useful for bio-filtration and gas separation, such as for controlling molecular transport in immunoisolation applications. Furthermore, AAO can also be formed to different geometric shapes. Yue [17] reported tubular anodic aluminum oxide (AAO) membranes with various geometries, including circle, square, and triangle, were fabricated using the "external anodizing" method. Adjust the membrane thickness through the second-step anodization time (10 to 28 h) was demonstrated to be feasible for avoiding the longitudinal cracks in various tubular AAO membranes.

In our previous research, we produced different nanostructures using high-quality AAO films [18–21]. However, the mass production of tubular AAO remains challenging. In this paper, based on our previous experience in AAO producing, a new technique is proposed for the batch production of tubular AAO films, and such batch production is an important factor in materials engineering [22].

2. Experimental Procedures

Tubular AAO films with an average pore size of about 60 nm and the film thickness of 40 μm were fabricated from 6061 aluminum alloy (Mg: 0.8~1.2%, Si: 0.4~0.8%, Cu: 0.15~0.40%, Al: base) tubes with a diameter of 20 mm by anodization processes. The experimental chemicals that were used in this experiment were of reagent grade without further purification, including perchloric acid (HClO$_4$), ethylene glycol butyl ether (CH$_3$(CH$_2$)$_3$OCH$_2$CH$_2$OH), ethanol (C$_2$H$_6$O), oxalic acid (C$_2$H$_2$O$_4$), chromium trioxide (CrO$_3$), phosphoric acid (H$_3$PO$_4$), cupric chloride (CuCl$_2$), and hydrogen chloride (HCl). The experimental steps are as below: (a) stress relief (6061 Al, 300 °C, 1 h); (b) mechanical grinding on the outer surface of the Al tube (#2000 SiC paper); (c) electrolytic polishing (15 vol. % HClO$_4$ + 15 vol. % (CH$_3$(CH$_2$)$_3$OCH$_2$CH$_2$OH + 70 vol. % C$_2$H$_6$O, 15 °C, 42 V, 10 min); (d) 1st anodization (3 wt. % C$_2$H$_2$O$_4$, 10 °C, 40 V, 1 h); (e) removal of 1st anodization film (1.8 wt. % CrO$_3$ + 6 vol. % H$_3$PO$_4$ + 92 vol. % H$_2$O, 70 °C, 50 min); (f) 2nd anodization (3 wt. % C$_2$H$_2$O$_4$, 10 °C, 40 V, 5 h); (g) removal of Al substrate (12 wt. % CuCl$_2$ + 8 vol. % HCl + 80 vol. % H$_2$O, 25 °C, 14 min); (h) removal of barrier layer (5 vol. % H$_3$PO$_4$, 25 °C, 1 h); and, (i) pore widening (5 vol. % H$_3$PO$_4$, 25 °C, 45 min).

The AAO thickness between 30 to 100 µm were controlled by the anodization time from 3 to 24 h. The transmembrane pressure test on the tubular AAO films was carried out by purging Ar gas, and the exudation characteristic of the tubular AAO was tested with a red ink solution.

All of the experimental sample images were acquired using a digital camera (Nikon, COOLPIX A900, Japan). The microstructures of the fabricated samples were observed by a scanning electron microscope (SEM, JEOL 6500).

3. Results and Discussion

In order to quickly fabricate and reduce the cost of the tubular AAO process, a batch of aluminum tubes was set in the anodization mold. Silicone plugs were used to seal the bottoms of the tubes so that AAO formed on the outer surfaces of the aluminum tubes rather than on the inner surfaces. Based on our previous anodization process that is illustrated in Figure 1 [1], the solutions of each tank individually were (a) electro-polishing electrolyte (HClO$_4$ solution); (b) water; (c) anodization electrolyte (oxalic acid solution); (d) AAO removal solution (H$_3$PO$_4$ solution); (e) aluminum removal solution (CuCl$_2$ + HCl solution); and, (f) AAO pore widening solution (H$_3$PO$_4$ solution). The experimental parameters of solution temperature and applied voltage were controlled using a cooling/heating machine and programmable power supply, respectively. The aluminum tube, functioning as a working electrode, was in contact with a conducting copper sheet as the anode, and an aluminum sheet was in contact with the cathode as the counter.

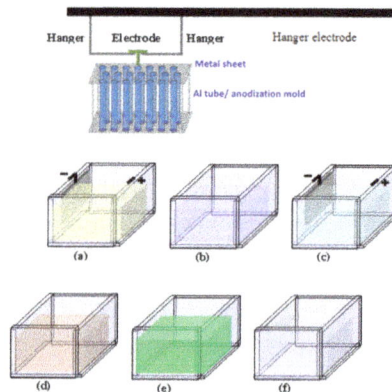

Figure 1. Schematic diagram of the producing flow and instrument setup with an overhead conductive rod for fabricating the tubular anodic aluminum oxide (AAO) template. The operating process is conducted in the order from (a–f), but after each step, the sample is rinsed in (b) the water bath. The solutions in the tanks are (a) electro-polishing electrolyte; (b) water; (c) anodization electrolyte; (d) AAO removal solution; (e) aluminum removal solution; and, (f) AAO pore widening solution.

Figure 2 shows the set up to remove Al substrate, including Al removal solution (CuCl$_2$ + HCl solution), pump, and pipes. After the removal solution was pumped into Al tube to etch Al for ten of minutes, translucent tubular AAO was obtained. The chemical displacement reaction between Al and Cu can be expressed as: 2 Al$_{(s)}$ + 3 CuCl$_{2(s)}$ → 3 Cu$_{(s)}$ + 2 AlCl$_{3(s)}$. By the way, HCl promotes the AlCl$_3$ formation, and the reaction is denoted as: 2 Al$_{(s)}$ + 6 H$^+_{(aq)}$ + 6 Cl$^-_{(aq)}$ → 2 AlCl$_{3(s)}$ + 3 H$_{2(g)}$. Because this is an exothermic reaction during Al-Cu displacement reaction the cooling for etching bath is critical. Figure 3 shows the detailed Al removing processing flow. Most of Al substrate can be removed after a 15 min removing process.

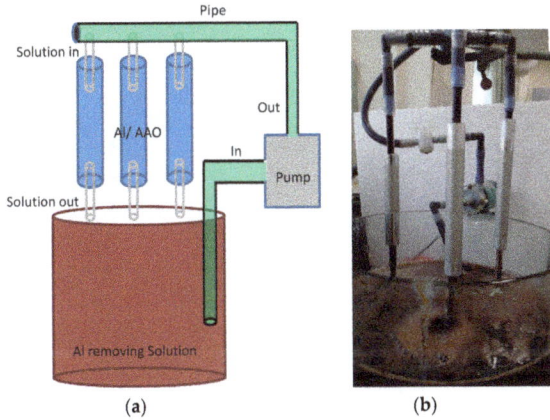

(a)　　　　　　　(b)

Figure 2. (a) Schematic diagram and (b) working situation of experimental setup for removing aluminum substrate from Al/AAO sample.

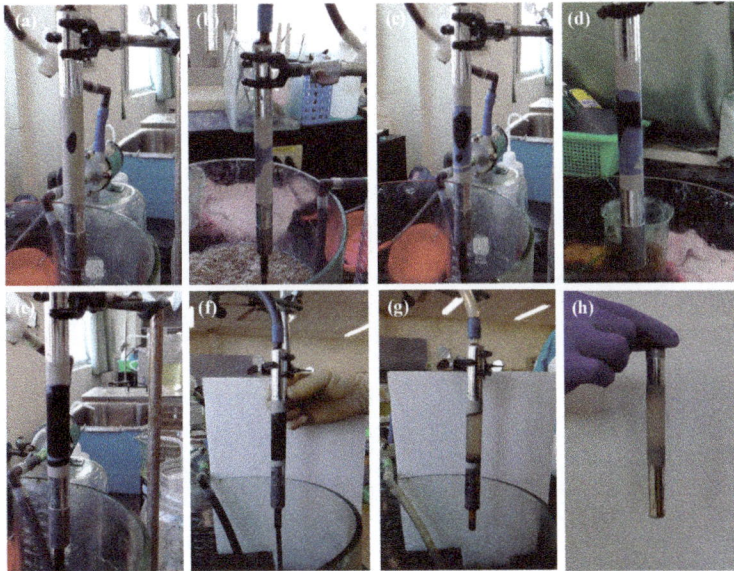

Figure 3. Optical images of tubular Al/AAO film after etching Al substrate for (a) 2 min; (b) 4 min; (c) 6 min; (d) 8 min; (e) 10 min; (f) 12 min; (g) 15 min; and, (h) the finish product cleaned with water.

Figure 4 shows the optical images of the initial 6061 tube and anodic tubular aluminum oxide made by the assistance of electrochemical and wet-etching process. It includes the following steps: (a) the aluminum tube that contains many scratches; (b) most of rough scratches on aluminum tube were removed after mechanical grinded; (c) a mirror-like aluminum surface was achieved after electrolytic polished; (d) a uniformity AAO film was formed on the Al surface through the first anodization process; (e) ordering patterns were formed on the Al tube after removing the first anodization film; (f) AAO film was formed on the ordering pattern Al surface through the second anodization; and, (g) a translucent AAO film was presented after removing Al substrate.

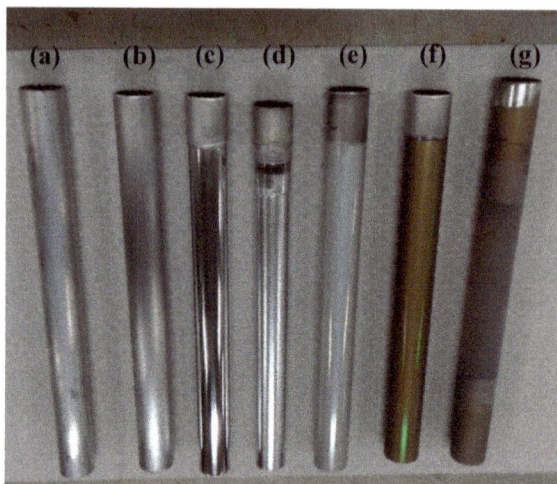

Figure 4. Step-by-step images of tubular AAO fabrication process: (**a**) initial 6061 Al tube and their outward appearances after each procedures, including (**b**) mechanical grinding; (**c**) electrolytic polishing; (**d**) the first anodization; (**e**) ordering pattern; and, (**f**) the second anodization; and, (**g**) the finished tubular AAO film with Al rings on both ends.

Figure 5 shows a schematic diagram of the tubular AAO micro-morphology. In the beginning, AAO film forms hexagonal channels because of the internal stress balance. Subsequently, the hexagonal channels are changed to a circular tube by anodization electrolyte etching. While the high purity Al tube is anodized, circular holes form on the hexagonal patterns of the anodic film. In Figure 5a, anodic aluminum oxide with regular pores on the Al tube surface was shown. Figure 5b is a lateral-view image of the anodic film, showing a straight channel with hemispherical closed barrier layer formation on the channel bottom. In Figure 5c, the aluminum substrate has been removed from the independent AAO film, and the anodic film presents a hemispherical closed barrier layer on the bottom. Because the closed barrier layer may affect the gas or liquid circulation in advanced applications, the barrier layer can be removed. In Figure 5d, the barrier layer is removed in a both-side open anodic channels. The anodic film structure is supported by a continuous wall, and both ends of the anodic tube are open to allow for air or liquid flow in the anodic tubes. However, in our experiment, 6061Al tubes were utilized as the raw material for anodization. AAO obtained from 6061Al has a worse microstructure than AAO prepared from pure Al.

Figure 6 shows photos of the tubular AAO film formed with 6061Al tubes. Figure 6a shows the raw material, the 6061 Al tube, which has a rough surface before mechanical and electrical polishing. Figure 6b shows that the 6061 Al tube presents a flat surface that is suitable for the formation of high-quality AAO after electro-polishing (EP) process. In Figure 6c, the AAO film on the EP Al further presents golden color by oxidized in oxalic acid solution with a DC voltage of at 40 V. Figure 6d shows tubular AAO film partially with the inner EP Al surface; therefore, the Al substrate could be entirely removed by dipping in CuCl$_2$ aqueous solution. The image shows that part of the Al substrate was removed, and the partially exposed AAO film is visible. In Figure 6e, the exposed AAO film with the Al rings on both sides further obtained after the Al substrate was totally removed. In future applications of tubular AAO, the Al rings on both sides can function as connectors. Figure 6f demonstrates that the red ink solution exuded through the AAO nano-pores.

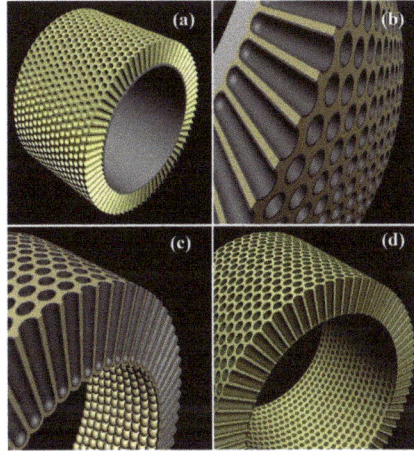

Figure 5. Schematic diagram of tubular AAO micro-morphology; (**a**) AAO film on the aluminum substrate and its (**b**) enlarged image; AAO film (**c**) without aluminum substrate; and, (**d**) barrier layer.

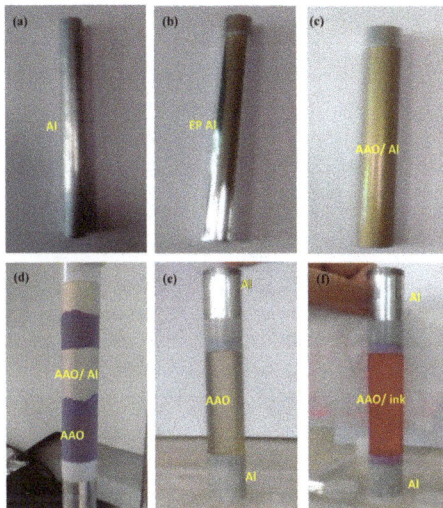

Figure 6. Photos of tubular AAO film formed from 6061 Al tubes: (**a**) raw 6061 Al tube material; (**b**) electro-polished (EP) 6061 Al; (**c**) AAO formation on the 6061 Al tube; (**d**) partially removed 6061 Al substrate from AAO film; (**e**) tubular AAO film without 6061 Al substrate; and, (**f**) red ink exudation through AAO film.

Figure 7 shows tubular AAO with Al connectors for the further characterizations of gas transmembrane pressure or liquid exudation tests. In Figure 7a, the schematic drawing of tubular AAO film with nano-channels structure is presented and both ends are connected with metal aluminum tubes. Figure 7b demonstrates the exploded view of a gas connector with tubular AAO, and the gas connectors on the tubular AAO in Figure 7c. The practical testing setup of gas connectors on the tubular AAO is shown in Figure 7d. For the pressure bearing testing of tubular AAO films, the barrier layer at the AAO bottom was kept, and then Ar gas was input into one of gas connector from the storage bottle, while the other end of gas connector was sealed. The pressure bearing capacity of tubular AAOs depended on their thickness. AAO film with a thickness of 100 μm can resist an argon pressure up to 8 atm; however, 30 μm AAO film can only withstand 3 atm of Ar gas. For the aeration test, AAO's barrier layer was removed, and Ar gas was injected into one gas connector (the other one was sealed). As aerating tubular AAO was put in the water, a lot of bubbles leaked out on the AAO surface.

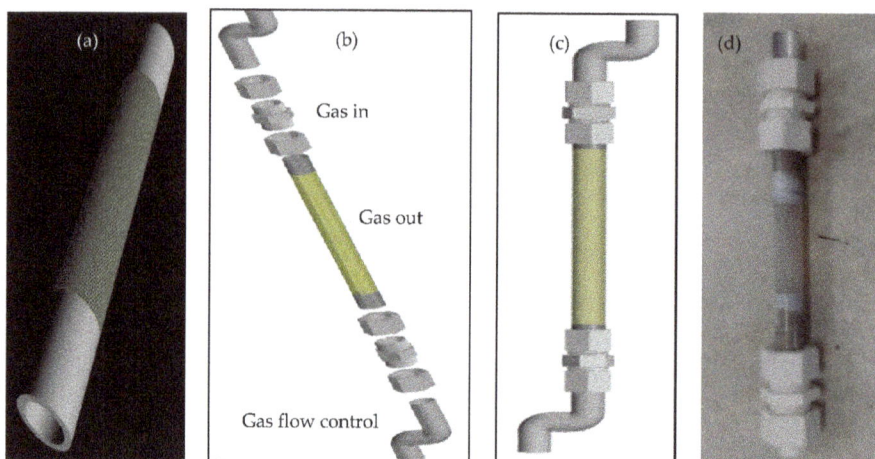

Figure 7. The schematic diagram of transmembrance pressure test on the tubular AAO; (**a**) tubular AAO; (**b**) exploded view of gas connector with tubular AAO; (**c**) gas connector with tubular AAO; and, (**d**) image of gas connector with tubular AAO.

Figure 8 presents the SEM images of tubular AAO anodized at 40 V. In Figure 8a, a barrier layer having an ordered compact structure was observed at the bottom of the AAO. Figure 8b shows the nano-pores after partially removing the barrier layer. In Figure 8c, the entire barrier layer has been removed, and the bottom view image reveals the nano-pore structure. In Figure 8d, the average pore dimension is about 60 nm and the pore density is about 10^8–10^9 pore/cm^2 in the AAO structure.

A complete tubular AAO process takes a long time, and the whole process includes (a) stress relief of Al tube (1 h); (b) mechanical grinding on the outer surface of the Al tube (1 h); (c) electrolytic polishing (10 min); (d) 1st anodization (1 h); (e) removal of 1st anodization film (50 min); (f) 2nd anodization (5 h); (g) removal of Al substrate (15 min); (h) removal of barrier layer (1 h); and, (i) pore widening (45 min). A complete tubular AAO process includes nine steps and takes around 10 h, at least. In order to increase the fabrication efficiency of tubular AAO a batch tubular AAO making is necessary. Figure 9 showed a batch tubular AAO process, including (a) a specified length of raw Al tubes is used; (b) a batch of mirror-like Al tubes surface formation after electrolytic polishing; (c) a batch of AAO production on the Al tubes through anodization; and, (d) a batch of tubular AAO formation after removing the retaining Al substrates. In our future work, high purity aluminum, such as 99.999% Al tubes will be used instead of 6061 Al for the AAO substrate. It is believed that the microstructure of the pore arrangement of AAO will thus be improved.

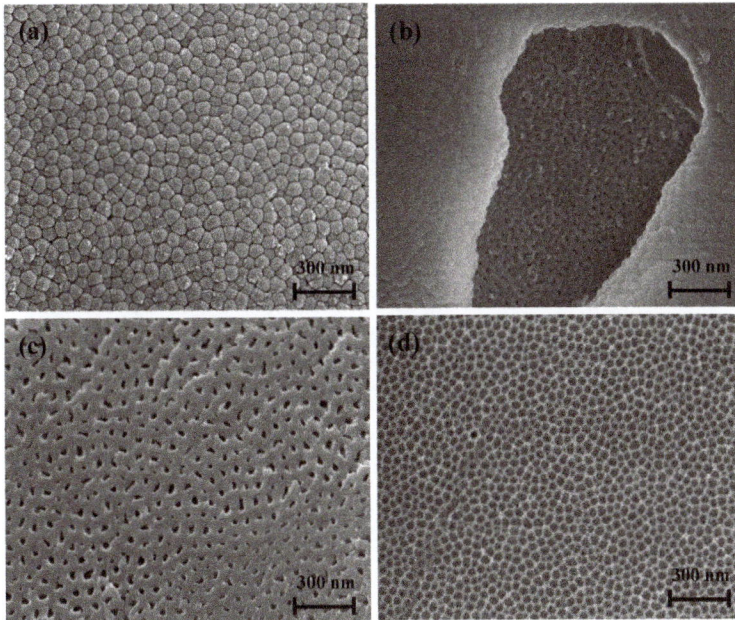

Figure 8. Scanning electron microscope (SEM) images of tubular AAO microstructure; (**a**) bottom view of barrier layer; (**b**) partially removed barrier layer; (**c**) bottom view without barrier layer; and, (**d**) top view of nano-pores.

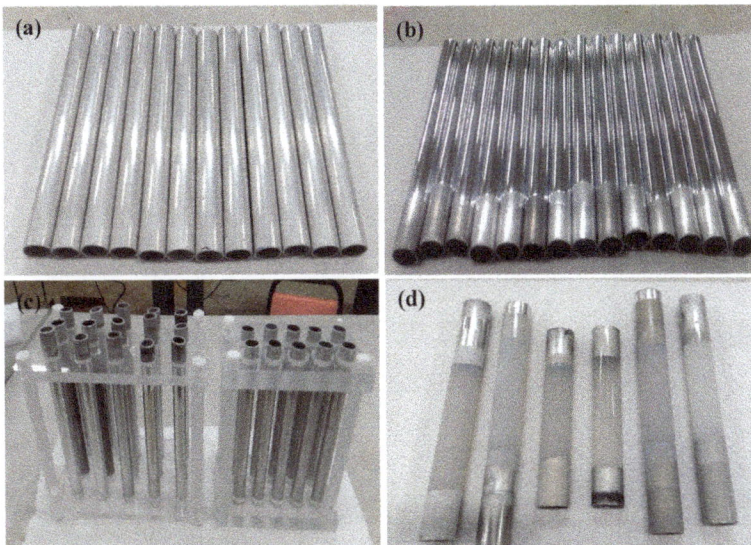

Figure 9. Batch tubular AAO process (**a**) 6061 Al tubes; (**b**) electrolytic polished 6061 Al tubes; (**c**) anodized 6061 Al tubes; and, (**d**) tubular AAO films.

4. Conclusions

A new and simple method was developed, which enables a cost-effective approach for the fabrication of a batch of tubular AAO templates. According to the SEM images, the pore arrangement of AAO is far less dense than the hexagonal pore structure. Based on the tubular AAO process, high-purity (99.999%) Al can be used to synthesize AAOs with pore diameters that are smaller than 10 nm in H_2SO_4 electrolyte at a lower applied voltage (18 V) and pore diameters larger than 400 nm in H_3PO_4 electrolyte at high applied voltage (200 V).

The applications of tubular AAO films were also demonstrated via Ar gas bearing test and aeration test. Tubular AAO films with a thickness of 100 μm can resist a gas pressure of up to 8 atm, and the resistance was still as high as 3 atm, even the film thickness reduced to 30 μm. With a further perforation treatment, tubular AAO films became fluid-permeable. According to the aeration and ink exudation tests, both Ar gas and red ink could past through nano-structured AAO films slowly, suggesting to a potential candidate for in filtering, dialysis, and gas diffuser applications.

Author Contributions: S.H.C. and C.C.C. conceived and designed the experiments; C.W.H. performed the experiments; S.H.C. and Y.-J.C. analyzed the data; Z.L. contributed verifications of results; C.C.C. wrote the paper.

Funding: This research received no external funding.

Acknowledgments: This work was financially supported by the Ministry of Science and Technology Taiwan, under the Contract No. 106-2221-E-239-023.

Conflicts of Interest: The authors declare no conflict of interest.

References

1. Chen, C.C.; Fang, D.; Luo, Z.P. Fabrication and Characterization of Highly-Ordered Valve-Metal Oxide Nanotubes and Their Derivative Nanostructures. *Rev. Nanosci. Nanotechnol.* **2012**, *1*, 229–256. [CrossRef]
2. Ide, S.; Capraz, Ö.Ö.; Shrotriya, P.; Hebert, K.R. Oxide Microstructural Changes Accompanying Pore Formation during Anodic Oxidation of Aluminum. *Electrochim. Acta* **2017**, *232*, 303–309. [CrossRef]
3. Jani, A.M.M.; Losic, D.; Voelcker, N.H. Nanoporous anodic aluminium oxide: Advances in surface engineering and emerging applications. *Prog. Mater. Sci.* **2013**, *58*, 636–704. [CrossRef]
4. Wang, Y.; Kaur, G.; Chen, Y.; Santos, A.; Losic, D.; Evdokiou, A. Bioinert anodic alumina nanotubes for targeting of endoplasmic reticulum stress and autophagic signaling: A combinatorial nanotube-based drug delivery system for enhancing cancer therapy. *ACS Appl. Mater. Interfaces* **2015**, *7*, 27140–27151. [CrossRef] [PubMed]
5. Nong, Y.L.; Qiao, N.N.; Deng, T.H.; Pan, Z.N.; Liang, Y. Solid sheet of anodic aluminium oxide supported palladium catalyst for Suzuki coupling reactions. *Catal. Commun.* **2017**, *100*, 139–143. [CrossRef]
6. Park, I.J.; Choi, S.R.; Kim, J.G. Aluminum anode for aluminum-air battery—Part II: Influence of In addition on the electrochemical characteristics of Al-Zn alloy in alkaline solution. *J. Power Sour.* **2017**, *357*, 47–55. [CrossRef]
7. Spooner, R.C. The Anodic Treatment of Aluminum in Sulfuric Acid Solutions. *J. Electrochem. Soc.* **1955**, *102*, 156–162. [CrossRef]
8. Akahori, H. Electron Microscopic Study of Growing Mechanism of Aluminum Anodic Oxide Film. *J. Electron Microsci.* **1961**, *10*, 175–185.
9. Mardilovich, P.P.; Govyadinov, A.N. New and modified anodic alumina membranes Part I. Thermotreatment of anodic alumina membranes. *J. Membr. Sci.* **1995**, *98*, 131–142. [CrossRef]
10. Altuntas, S.; Buyukserin, F. Fabrication and characterization of conductive anodic aluminum oxide substrates. *Appl. Surface Sci.* **2014**, *318*, 290–296. [CrossRef]
11. Belwalkar, A.; Grasing, E.; Geertruyden, W.V.; Huang, Z.; Misiolek, W.Z. Effect of processing parameters on pore structure and thickness of anodic aluminum oxide (AAO) tubular membranes. *J. Membr. Sci.* **2008**, *319*, 192–198. [CrossRef] [PubMed]
12. Gong, D.; Yadavalli, V.; Paulose, M.; Pishko, M.; Grimes, C.A. Controlled Molecular Release Using Nanoporous Alumina Capsules. *Biomed. Microdevices* **2003**, *5*, 75–80. [CrossRef]

13. Kasi1, A.K.; Kasi1, J.K.; Hasan, M.; Afzulpurkar, N.; Pratontep, S.; Porntheeraphat, S.; Pankiew, A. Fabrication of low cost anodic aluminum oxide (AAO) tubular membrane and their application for hemodialysis. *Adv. Mater. Res.* **2012**, *550–553*, 2040–2045. [CrossRef]

14. Kasi, J.K.; Kasi, A.K.; Bokhari, M.; Sohail, M. Characterization of Cracks in Tubular Anodic Aluminum Oxide Membrane. *Am. J. Condens. Matter Phys.* **2016**, *6*, 36–40.

15. Itoh, N.; Tomura, N.; Tsuji, T.; Hongo, M. Strengthened porous alumina membrane tube prepared by means of internal anodic oxidation. *Microporous Mesoporous Mater.* **1998**, *20*, 333–337. [CrossRef]

16. Grimes, C.A.; Gong, D.W. Tubular Filter with Branched Nanoporous Membrane Integrated with a Support and Method of Producing Same. U.S. Patent 20030047505 A1, 13 March 2003.

17. Yue, S.; Zhang, Y.; Du, J. Preparation of anodic aluminum oxide tubular membranes with various Geometries. *Mater. Chem. Phys.* **2011**, *128*, 187–190. [CrossRef]

18. Chen, C.C.; Chang, S.F.; Luo, Z.P. Anodic-aluminum-oxide template assisted fabrication of cesium iodide (CsI) scintillator Nanowires. *Mater. Lett.* **2013**, *112*, 190–193. [CrossRef]

19. Chen, C.C.; Bisrat, Y.; Luo, Z.P.; Schaak, R.E.; Chao, C.G.; Lagoudas, D.C. Fabrication of Single Crystal tin Nanowires by Hydraulic Pressure Injection. *Nanotechnology* **2006**, *17*, 367–374. [CrossRef]

20. Chen, P.C.; Hsieh, S.J.; Chen, C.C.; Zou, J. The Microstructure and Capacitance Characterizations of Anodic Titanium Based Alloy Oxide Nanotube. *J. Nanomater.* **2013**, *2013*. [CrossRef]

21. Chen, M.S.; Fan, F.Y.; Lin, C.K.; Chen, C.C. Enhanced Diffusion Bonding Between High Purity Aluminum and 6061 Aluminum by Electrolytic Polishing Assistance. *Int. J. Electrochem. Sci.* **2016**, *11*, 4922–4929. [CrossRef]

22. Chen, C.C.; Chang, S.F.; Li, L.L.; Chou, C.Y.; Yang, H.W. A Tubular Solar Cell. TW Patent I590502, 14 November 2016.

MDPI

St. Alban-Anlage 66

4052 Basel

Switzerland

Tel. +41 61 683 77 34

Fax +41 61 302 89 18

www.mdpi.com

Applied Sciences Editorial Office

E-mail: applsci@mdpi.com

www.mdpi.com/journal/applsci

www.ingramcontent.com/pod-product-compliance
Lightning Source LLC
Chambersburg PA
CBHW051908210326
41597CB00033B/6076